TOPICS ON REAL AND COMPLEX SINGULARITIES

TOPICS ON REAL AND COMPLEX SINGULARITIES

Proceedings of the 4th Japanese–Australian Workshop (JARCS4)

Kobe, Japan   22 – 25 November 2011

Editors

Satoshi Koike
Hyogo University of Teacher Education, Japan

Toshizumi Fukui
Saitama University, Japan

Laurentiu Paunescu
University of Sydney, Australia

Adam Harris
University of New England, Australia

Alexander Isaev
Australian National University, Australia

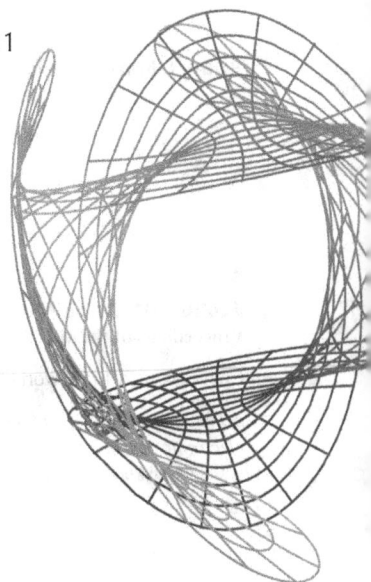

World Scientific

NEW JERSEY · LONDON · SINGAPORE · BEIJING · SHANGHAI · HONG KONG · TAIPEI · CHENNAI

Published by

World Scientific Publishing Co. Pte. Ltd.
5 Toh Tuck Link, Singapore 596224
USA office: 27 Warren Street, Suite 401-402, Hackensack, NJ 07601
UK office: 57 Shelton Street, Covent Garden, London WC2H 9HE

British Library Cataloguing-in-Publication Data
A catalogue record for this book is available from the British Library.

TOPICS ON REAL AND COMPLEX SINGULARITIES
Proceedings of the 4th Japanese–Australian Workshop (JARCS4)

ISBN 978-981-4596-03-9

Preface

The fourth Japanese-Australian Workshop on Real and Complex Singularities (JARCS4 KOBE 2011) was held at the Kobe Satellite of Hyogo University of Teacher Education during the period 22-25 November, 2011. There were 31 participants from Australia and Japan.

The Australian and Japanese Singularity Theory specialists have built up a strong research relationship in the past three decades. For instance, the blow-analytic theory introduced by Tzee-Char Kuo in Australia has been intensively developed in Japan. In addition, a lot of joint works by researchers of both countries have been established in several topics related to real and complex singularities.

The present volume consists mainly of the texts of the invited talks of the workshop. Some of them are joint works of Australians and Japanese.

This volume contains original articles on real and complex singularities, topology of differentiable maps, openings of differentiable map-germs, the relationship between free divisors and holonomic systems, effective computational method of invariants of singularities, the application of singularity theory to differential geometry, the deformation theory of CR structures and differential equations with singular points. In these articles some important new notions for characterizations of singularities are introduced, and several new results are presented. New approaches to classical topics and new computational methods of singularities are also presented.

We would like to thank the contributors for their cooperation. All the articles in this volume have been very carefully refereed. We would also like to thank the referees for taking time to read and give a lot of comments again and again.

The workshop was supported by the Grant-in-Aid for Scientific Researches

No. 23244008 (Investigator: Osamu Saeki),
No. 22340030 (Investigator: Goo Ishikawa),
No. 23540099 (Investigator: Kimio Miyajima),
No. 21540054 (Investigator: Shuzo Izumi),
No. 23540087 (Investigator: Satoshi Koike)

of the Ministry of Education, Culture, Sports, Science and Technology (MEXT) of Japan, and Hyogo University of Teacher Education. We acknowledge the support.

18 September 2013

Satoshi Koike

Organizing Committees

of the fourth Japanese-Australian Workshop on
Real and Complex Singularities

Satoshi Koike (Chairman) – Hyogo University of Teacher Education,
 Japan
Toshizumi Fukui – Saitama University, Japan
Laurentiu Paunescu – University of Sydney, Australia
Adam Harris – University of New England, Australia
Alexander Isaev – Australian National University, Australia

List of Participants

Takao Akahori (Hyogo University, Japan)

Kana Ando (Chiba University, Japan)

Takuo Fukuda (Nihon University, Japan)

Toshizumi Fukui (Saitama University, Japan)

Hirofumi Hagi (Hyogo University of Teacher Education, Japan)

Adam Harris (University of New England, Australia)

Jonathan Hillman (University of Sydney, Australia)

Goo Ishikawa (Hokkaido University, Japan)

Shuzo Izumi (Kinki University, Japan)

Shigeyasu Kamiya (Okayama University of Science, Japan)

Yoshihiro Kawazoe (Hyogo University of Teacher Education, Japan)

Yumiko Kitagawa (Oita National College of Technology, Japan)

Naoki Kitazawa (Tokyo Institute of Technology, Japan)

Mahito Kobayashi (Akita University, Japan)

Satoshi Koike (Hyogo University of Teacher Education, Japan)

Yutaka Matsui (Kinki University, Japan)

Kimio Miyajima (Kagoshima University, Japan)

Masayuki Nishioka (Kyushu University, Japan)

Masao Ogawa (Hyogo University of Teacher Education, Japan)

Mutsuo Oka (Tokyo University of Science, Japan)

Tomohiro Okuma (Yamagata University, Japan)

Laurentiu Paunescu (University of Sydney, Australia)

Osamu Saeki (Kyushu University, Japan)

Jiro Sekiguchi (Tokyo University of Agriculture and Technology, Japan)

Ryuhei Shigematsu (Hyogo University of Teacher Education, Japan)

Masahiro Shiota (Nagoya University, Japan)

Shinichi Tajima (Tsukuba University, Japan)

Kiyoshi Takeuchi (Tsukuba University, Japan)

Takahiro Yamamoto (Kyushu Sangyo University, Japan)

Wataru Yukuno (Hokkaido University, Japan)

Ruibin Zhang (University of Sydney, Australia)

Contents

On the CR Hamiltonian flows and CR Yamabe problem

Takao Akahori

Hyogo University, Kobe, Hyogo, 651-2197, Japan
akahorit@sci.u-hyogo.ac.jp

CR Yamabe problem was treated by John M. Lee and several people. Their setting is that: by fixing the CR structure, they try to find a suitable contact form. Our new approach differs from theirs. We fix the contact form and change CR structures. Our approach is in progress.

Keywords: CR Yamabe problem, deformation theory of CR structures
AMS classification numbers: 32V20

The deformation theory of CR structures is initiated by Kuranishi, and developed by several authors and in several ways (for example, Miyajima, Lee, Cheng, and myself). Especially, in [A4], the notion of CR-Hamiltonian flows is found. Let $\{(M, {}^0T''), \theta\}$ be a CR structure with a contact form, embedded in a complex manifold N as a real hypersurface. Then, we have the Cartan connection, and have the scalar curvature R. Related with the Einstein Kaehler metric, CR Yamabe problem is of interest. This problem is already treated by J. M. Lee. His approach is that; for any given CR structure with a contact form $\{(M, {}^0T''), \theta\}$, is there a real valued C^∞ non vanishing function g, which satisfies $\{(M, {}^0T''), g\theta\}$ that has a constant scalar curvature? Our approach differs from his. We fix the contact form and change the CR structure by the CR Hamiltonian flows. We set our approach. We assume that $\{(M, {}^0T''), \theta\}$ has a constant scalar curvature, and consider the deformation of the contact structure, $\{(M, {}^0T''), \frac{1}{\rho^2}\theta\}$, where ρ is a C^∞ function, close to 1. Then, we cannot expect that our CR structure with the corresponding contact structure admits the constant scalar curvature. Let $R(\rho)$ denote its scalar curvature. Our problem is that: if ρ is close enough to 1, is it possible to show that: there is a real valued C^∞ function h which satisfies the scalar curvature of $\{(M, {}^{\phi(\sqrt{-1}h)}T''), \frac{1}{\rho^2}\theta\}$ is constant? Here $(M, {}^{\phi(\sqrt{-1}h)}T'')$ means the CR structure, induced by the CR Hamiltonian flow, generated by $\sqrt{-1}h$. We see the model case, the Heisenberg structure with real dimension 3. And let the contact structure be the standard one. Then, all curvature tensors vanish, and so the scalar curvature vanishes. But if we adopt $\frac{1}{\rho^2}\theta$ as a contact form, where θ is the standard contact form and ρ is a C^∞ function close to 1, then the scalar curvature may not vanish. For this structure, by changing the CR structure by the CR-Hamiltonian flow (the contact form is $\frac{1}{\rho^2}\theta$ and this is fixed), we would like to obtain the constant scalar curvature (in this case, 0 scalar curvature). This problem becomes an existence problem of the solution of the non-linear partial differential equation. For the linear part of this non-linear partial differential equation, we show the sub-ellipticity (the main theorem). This suggests that our problem would be solved affirmatively. And also, it might suggest the deformation of contact structures is something like a dual space with the deformation of CR structures, induced by the CR Hamiltonian flow, via the scalar curvature. While for our original problem, we have to see the non-linear part more carefully. In a future paper, with a wide scope, we discuss this problem. Finally, the author has to mention that this problem is inspired by the Kuranishi's work for the Cartan connection.

1. Deformation theory of CR structures

Let M a real $2n - 1$ dimensional C^∞ manifold. And we assume that our M is embedded in a complex manifold N. Let $^0T''$ be a sub bundle of the complexfied tangent vector bundle of M. Now, pair $(M, {}^0T'')$ is called a CR structure iff the following two conditions are satisfied.

$$(1) \ ^0T'' \cap \overline{^0T''} = 0, \ dim_{\mathbf{C}} \frac{C \otimes TM}{^0T'' + \overline{^0T''}} = 1,$$

$$(2) \ [\Gamma(M, {}^0T''), \Gamma(M, {}^0T'')] \subset \Gamma(M, {}^0T'')$$

A pair (M, θ) is called a contact structure if a real 1-form on M, θ, which satisfies the following condition:

$$\theta \wedge (d\theta)^{n-1} \neq 0 \text{ at any point of } M$$

is given. So a contact structure is nothing with a CR structure.

Now let (V, o) be an isolated singularity in a complex euclidean space $\mathbf{C}^N$. Let

$$M = V \cap S_\epsilon^{2N-1}(o)$$

where $S_\epsilon^{2N-1}(o)$ is a hypersphere, centered at the origin with radius ϵ.

Then, over M, naturally, a CR structure

$$^0T'' := C \otimes T(V - o) \cap T'' \mathbf{C}^N$$

is induced. In this case, our CR structure is strongly pseudo convex. Let

$$r := \sum_{i=1}^{N} |z_i|^2 - \epsilon$$

and

$$\theta := \sqrt{-1}\partial r.$$

Then, our θ is a real 1-form on M and (M, θ) is a contact structure. We note that if g is any positive C^∞ function on M, $(M, g\theta)$ is a different contact structure. In this sense, a contact structure is a completely different notion from a CR structure. Now we recall the deformation equation of CR structures. Let $(M, {}^0T'')$ be a strongly pseudo convex CR structure. We fix a contact structure which is compatible with our CR structure. Here "compatible" means that:

$$\theta \mid_{{}^0T'' + {}^0T'} = 0$$

where ${}^0T' = \overline{{}^0T''}$.

By using this θ, we determine a real vector ξ on M by

$$\begin{cases} \theta(\xi) = 1 \\ d\theta(W, \xi) = 0, \ W \in {}^0T' \end{cases}$$

Now by using this ξ, we give a C^∞ vector bundle decomposition of $C \otimes TM$.

$$C \otimes TM = {}^0T'' + C \otimes \xi + {}^0T'.$$

Set $T' := C \otimes \xi + {}^0T'$. Then, T' has the following meaning. Consider the following scheme.

$$T' \to C \otimes TM \hookrightarrow C \otimes T(V - o) \mid_M$$
$$\downarrow$$
$$T'(V - o) \mid_M$$

By comparing the dimension of vector bundles T' and $T'(V - o)$ the map: $T' \to T'(V - 0)$ is an isomorphism map. So our T' is a natural one.

Later on, we write the vector bundle decomposition of $C \otimes TM$ as follows.

$$C \otimes TM = {}^0T'' + T'. \tag{1.1}$$

With these preparations, we consider the deformation theory of CR structures of $(M, {}^0T'')$.

Proposition 1.1. *Almost CR manifold with finite distance from $(M, {}^0T'')$ is one to one correspondence to $\Gamma(M, \mathrm{Hom}({}^0T'', T'))$. The correspondence is that; $\phi \in \Gamma(M, \mathrm{Hom}({}^0T'', T')) = \Gamma(M, T' \otimes ({}^0T'')^*)$*

$$^\phi T'' := \{X' : X' = X + \phi(X), X \in {}^0T''\}$$

By using this correspondence, we see that when our almost CR structure is actually a CR structure.

Proposition 1.2. *Almost CR manifold $(M, {}^\phi T'')$ is really CR structure iff our ϕ satisfies the following non-linear partial differential equation.*

$$\overline{\partial}_{T'}^{(1)} \phi + R_2(\phi) + R_3(\phi) = 0. \tag{1.2}$$

We see $\overline{\partial}_{T'}^{(1)}$ more precisely. First, we set $\overline{\partial}_{T'} : \Gamma(M, T') \to \Gamma(M, T' \otimes ({}^0T'')^*)$ by; for $u \in \Gamma(M, T')$, $\overline{\partial}_{T'}u(Y) = [Y, u]_{T'}$, $Y \in {}^0T''$. And as for scalar valued differential forms, we introduce $\overline{\partial}_{T'}^{(i)}$. Then, the case $i = 1$ is exactly the same as above operators. Now by the integrability condition, we have a differential complex.

Standard deformation complex

$$0 \longrightarrow \Gamma(M, T') \xrightarrow{\overline{\partial}_{T'}} \Gamma(M, T' \otimes ({}^0T'')^*) \xrightarrow{\overline{\partial}_{T'}^{(1)}} \Gamma(M, T' \otimes \wedge^2({}^0T'')^*)...$$

This differential complex $(M, {}^0T'')$ can be defined without strongly pseudo convexity. Especially, if our CR structure is strongly convexity and M is compact, our complex is sub-elliptic. and in the case $\dim_{\mathbf{R}} M = 2n - 1 \geq 5$,

$$\dim_{\mathbf{C}} H^1(M, T') := \mathrm{Ker}\,\overline{\partial}_{T'}^{(1)} / \mathrm{Im}\,\overline{\partial}_{T'} < +\infty$$

However, for the deformation theory, the standard complex is not enough, and so we introduce the new deformation complex.

New deformation complex

$$\mathcal{H} \xrightarrow{\overline{\partial}_{T'}} \Gamma(M, E_1) \xrightarrow{\overline{\partial}_{T'}^{(1)}} \Gamma(M, E_2)...$$

We have to explain notations. For $j = 1$,

$$\Gamma(M, E_1) = \{u : u \in \Gamma(M, {}^0T' \otimes \wedge^1({}^0T'')^*), (\overline{\partial}_{T'}^{(1)}u)_{C\xi \otimes \wedge^2({}^0T'')^*=0}\}.$$

And $j \geq 2$,

$$\Gamma(M, E_j) = \{v : v \in \Gamma(M, {}^0T' \otimes \wedge^j({}^0T'')^*), (\overline{\partial}_{T'}^{(j)}u)_{C\xi \otimes \wedge^{j+1}({}^0T'')^*=0}\}.$$

Here in the definition of $\Gamma(M, E_j)$, we note $^0T'$-valued (not T'-valued). And

$$\mathcal{H} = \{w : w \in \Gamma(M, T'),\ (\bar{\partial}_{T'}w)_{C\xi \otimes (^0T'')^*} = 0\}$$

This $\mathcal{H}$ is an infinite dimensional vector space (there is no vector bundles which satisfies ; $\mathcal{H}$ is the global sections).

We discuss this differential in more details. This differential complex comes from

$$\Gamma(M, \mathbf{C}) \xrightarrow{\ D\ } \Gamma(M, E_1) \xrightarrow{\ \bar{\partial}_{T'}^{(1)}\ } \Gamma(M, E_2)...$$

Here

$$g \in \Gamma(M, \mathbf{C}),\ \ Dg := \bar{\partial}_{T'}(g\xi + X_g) \in \Gamma(M, E_1).$$

We note that D is a second order partial differential operators.

We explain $g\xi + X_g$.

Definition 1.3 (CR Hamilton vector). *Let g be a C^∞ function on M. Let X_g be $^0T'$ valued section on M, defined by*

$$[X_g, Y]_{\mathbf{C}\xi} = (\bar{\partial}_b g)(Y)\xi,\ \text{for } Y \in {}^0T''$$

We call this complex vector $g\xi + X_g$ a CR Hamilton vector. And also we call g a generating function.

Proposition 1.4. *Let $g\xi + X_g$ be a CR Hamilton vector. Then, we have*

$$\bar{\partial}_{T'}(g\xi + X_g)\ is\ {}^0T' \otimes (^0T'')^* - valued.$$

more precisely, it holds that ; $g\xi + X_g \in \mathcal{H}$.

2. CR Hamiltonian flows

In our former paper (see [A4]), we introduced the notion of CR Hamiltonian flows. Actually, there are two types. For any C^∞ function g,

    (type 1) $g\xi + X_g$

This is the kernel of our new complex. Furthermore,

    (type 2) $g\xi - X_g$

This vector field preserves the contact form θ. In fact, for $Y \in {}^0T''$,

$$\mathcal{L}_{g\xi - X_g}\theta(Y) = d\theta(g\xi - X_g, Y) + d(\theta(g\xi - X_g))(Y)$$
$$= -d\theta(X_g, Y) + dg(Y)$$
$$= 0.$$

For any complex valued C^∞ function g, the existence of CR Hamiltonian flow is proved in the same way (in the former paper, the proof is given for type 2 vectors, but by the complete same way, we can obtain the same statement). Probably, more serious studies might be needed for these two type of vectors. Here, we see vectors which are of type 1.

Let g be a complex-valued C^∞ function on M. We assume that our M is embedded in a complex manifold N as a real hypersurface. Roughly speaking, $p \in M \to f_g(p) \in N$.

Here, in our case, for any real valued C^∞ function h, we solve that: for a sufficiently small positive number ϵ, for $-\epsilon < t < \epsilon$,

$$\begin{cases} \frac{d}{dt} f_{\sqrt{-1}h}(t) = (\sqrt{-1}h\xi + (Lh)\overline{L}) f_{\sqrt{-1}ht} \\ f_{\sqrt{-1}h}(0) = \text{identity map} \end{cases}$$

It is not clear that we can put $t = 1$. Instead, if $\mid h \mid$ is chosen sufficiently small, then we have

$$z^i \cdot f_{\sqrt{-1}h} = (\sqrt{-1}h\xi + (Lh)\overline{L})z^i + O(\text{higher order term of } h).$$

$j = 1, \ldots, n$ and z^j are complex coordinates of $\mathbf{C}^n$ And the corresponding deformation is

$$\phi(\sqrt{-1}h) = \overline{\partial}_{T'}((\sqrt{-1}h)\xi + X_{\sqrt{-1}h}) + O(h^2).$$

3. The scalar curvature

As for a complex valued C^∞ fuction g (in Sect. 2), we take $\sqrt{-1}h$, where h is a real valued C^∞ function. For this complex valued function, we consider the Hamiltonian flow, and $\phi(\sqrt{-1}h)$ denotes the corresponding form by the induced CR structure by this Hamiltonian flow. As mentioned,

$$\phi(\sqrt{-1}h) = \overline{\partial}_{T'}((\sqrt{-1}h)\xi + X_{\sqrt{-1}h}) + O(h^2).$$

We consider the following map.

$$h \in C^\infty(\mathbf{R}) \to \text{Scalar curvature of the CR structure } \{(M, {}^{\phi(\sqrt{-1}h)}T''), \theta\}.$$

Here, we note that the contact form θ is fixed. By following Lee-Jerison (see [J-L1]), we recall the notion of scalar curvature. First, we set the Tanaka-Webster connection,

$$\nabla : \Gamma(M, {}^0T') \to \Gamma(M, {}^0T' \otimes (C \otimes TM)^*)$$

satisfying:

(1) compatible with the CR structure (this means that $\nabla_{\overline{W}}$, for $\overline{W} \in {}^0T''$, is determined),
(2) preserving the Levi metric (so ∇_W, for $W \in {}^0T'$ is determined),

(3) and for the supplement vector, we set $\nabla_\xi W = [\xi, W]_{^0T'}$ for $W \in \Gamma(M, {}^0T')$, where $[\xi, W]_{^0T'}$ means the ${}^0T'$ part of $[\xi, W]$.

With these setting, we introduce the scalar curvature. Let $\{W_1, ., W_{n-1}\}$ an orthonormal system of bases of ${}^0T'$ with respect to the Levi metric, determined by the contact form θ. By these vectors, we determine a 1 form $\omega_\alpha{}^\beta$ by

$$\nabla W_\alpha = \sum_\beta \omega_\alpha{}^\beta \otimes W_\beta.$$

The scalar curvature is determined by

$$\sum_{\alpha, \rho} \{(d\omega_\alpha{}^\alpha - \sum_\gamma \omega_\alpha{}^\gamma \wedge \omega_\gamma{}^\alpha)(W_\rho, \overline{W_\rho})\}$$

Needless to say, it does not depend on the choice of the orthonormal system.

4. The model case 1 (changing the contact structure)

We see our Yamabe problem for the case Heisenberg structure with real dimension 3. In $\mathbf{C}^2$, we consider

$$\{(z_1, z_2) : (z_1, z_2) \in \mathbf{C}^2, \operatorname{Im} z_2 = \frac{1}{2}|z_1|^2\}.$$

And we set $L = \frac{\partial}{\partial \bar{z}_1} - \sqrt{-1}z_1 \frac{\partial}{\partial \bar{z}_2}$. ${}^0T''$ is generated by L. Let $\theta = 2\sqrt{-1}\partial r$, where $r = \operatorname{Im} z_2 - \frac{1}{2}|z_1|^2$. In this case, all curvature tensors $\omega_\alpha{}^\beta$ vanish. We change the contact form. For the contact form, we adopt $\frac{1}{\rho^2}\theta$. Here, ρ is a non vanishing real valued C^∞ function. In this case, some curvature tensor $\omega_\alpha{}^\beta$ may not vanish. In fact, our L is no more the unit vector. $L' := \rho L$ is the unit vector. Our connection is compatible with the CR structure. So,

$$\begin{aligned}
\nabla_{L'}\overline{L'} &= \nabla_{\rho L}\rho\overline{L} \\
&= [\rho L, \rho\overline{L}]_{^0T'} \\
&= \frac{1}{\rho}(L'\rho)\overline{L'}
\end{aligned}$$

We see $\nabla_{\overline{L'}}\overline{L'}$. Our connection preserves the metric. So,

$$\overline{L'}\langle \overline{L'}, \overline{L'} \rangle = \langle \nabla_{\overline{L'}}\overline{L'}, \overline{L'} \rangle + \langle \overline{L'}, \nabla_{L'}\overline{L'} \rangle$$

Hence

$$0 = \langle \nabla_{\overline{L'}}\overline{L'}, \overline{L'} \rangle + \langle \overline{L'}, \nabla_{L'}\overline{L'} \rangle$$

So,

$$\nabla_{\overline{L'}}\overline{L'} = -\frac{1}{\rho}(\overline{L'}\rho)\overline{L'}.$$

Set $\nabla \overline{L'} = \omega_1{}^1 \otimes \overline{L'}$. And we determine the connection 1-form $\omega_1{}^1$. Then,

$$\omega_1{}^1(L') = \frac{1}{\rho}(L'\rho), \qquad (4.3)$$

$$\omega_1{}^1(\overline{L'}) = -\frac{1}{\rho}(\overline{L'}\rho). \qquad (4.4)$$

By the definition, the scalar curvature, here we write it by $R(\rho)$, is

$$d\omega_1{}^1(\overline{L'}, L').$$

We compute the principal part of this $R(\rho)$.

$$\begin{aligned}
d\omega_1{}^1(\overline{L'}, L') &= \overline{L'}\omega_1{}^1(L') - L'\omega_1{}^1(\overline{L'}) - d\omega_1{}^1([\overline{L'}, L']) \\
&= \rho(\overline{L}L\rho + L\overline{L}\rho) + \text{lower order derivatives} \\
&\simeq (\overline{L}L\rho + L\overline{L}\rho) + \text{lower order derivatives}
\end{aligned}$$

Here $\simeq$ means "mod $(1-\rho)^2$". This is a highly non-trivial term, and so may not vanish.

5. The model case 2 (changing the CR structure)

We displace this hypersurface by the Hamiltonian flow, generated by $\sqrt{-1}h$, where h is a real valued C^∞ function. $\omega_\alpha{}^\beta(\rho, \sqrt{-1}h)$ denotes the corresponding curvature for this displacement (actually, in our case, there is only one component), $\omega_1{}^1(\rho, \sqrt{-1}h)$. We would like to determine the principal part of the liner term of $\omega_1{}^1(\rho, \sqrt{-1}h)$ with respect to h. In this case, $\dim_{\mathbf{C}}{}^0T'' = 1$. So, our CR structure is generated by L.

$$[\overline{L}, L]_\xi = \sqrt{-1}\xi,$$

where $\xi = \frac{\partial}{\partial z_2} + \frac{\partial}{\partial \overline{z}_2}$. So, $X_{\sqrt{-1}h} = (Lh)\overline{L}$. That is to say,

$$Z_{\sqrt{-1}h} = \sqrt{-1}h\xi + (Lh)\overline{L}.$$

Now we consider the deformation of this CR structure. The case CR Hamiltonian preserves ${}^0T' + {}^0T''$ (ξ part does not appear). This is not a trivial result. But, infinitesimally, it is obvious. And we are looking at the linear term. And so we can assume that our deformation is of type;

$$L'^\psi = L' + \psi \cdot \overline{L'}.$$

$$^\psi T'' = \{\text{complex vectors generated by } L'^\psi\}$$

And consider the scalar curvature for $(M, {}^\psi T'')$. Here, ψ is a C^∞ function and L' is the same as in Section 4. As mentioned, our L'^ψ is still the unit vector with respect to mod $|\psi|^2$. And

$$\begin{aligned}
\nabla_{L'^\psi} \overline{L'^\psi} &= \nabla_{\rho L^\psi} \rho \overline{L}^\psi \\
&= [\rho L^\psi, \rho \overline{L}^\psi]_{\psi T'} \\
&= [\rho(L' + \psi \overline{L'}), \rho(\overline{L'} + \overline{\psi} L')]_{\psi T'} \\
&\simeq -(\overline{L'^\psi}\psi)\overline{L'^\psi} + \text{lower order derivatives} \\
&\simeq -(\overline{L'}\psi)\overline{L'^\psi} + \text{lower order derivatives}
\end{aligned}$$

Here $\simeq$ means mod $|\psi|^2$.

We see $\nabla_{\overline{L'^\psi}} \overline{L'^\psi}$. Similarly,

$$\nabla_{\overline{L'^\psi}} \overline{L'^\psi} \simeq (L'\overline{\psi})\overline{L'^\psi}$$

Let $\omega_1{}^1(\rho, \psi)$ denote the corresponding the connection 1-form. By these computations,

$$\omega_1{}^1(\rho, \psi)(L'^\psi) \simeq -(\overline{L'}\psi), \tag{5.5}$$
$$\omega_1{}^1(\rho, \psi)(\overline{L'^\psi}) \simeq L'\overline{\psi}. \tag{5.6}$$

(Compare with (4.3), (4.4).) While, ξ-direction, we have the following lemma.

Lemma 5.1. *The principal part of the linear term of $\omega_1{}^1(\rho, \psi)(\xi)$ with respect to ψ, is zero.*

Proof. By the definition of the connection 1-form,

$$\omega_1{}^1(\rho, \psi)(\xi)(\overline{L} + \overline{\psi} L) = [\xi, \overline{L} + \overline{\psi} L]_{\psi T'}.$$

Here $[\xi, \overline{L} + \overline{\psi} L]_{\psi T'}$ means the $^\psi T'$ part of $[\xi, \overline{L} + \overline{\psi} L]$ according to the C^∞ vector bundle decomposition

$$\mathbf{C} \otimes TM = {}^\psi T'' + {}^\psi T' + \mathbf{C} \otimes \xi$$

While,

$$[\xi, \overline{L} + \overline{\psi} L] = (\xi \overline{\psi}) L.$$

We determine $^\psi T'$ part. There are C^∞ functions A,B satisfying:

$$(\xi \overline{\psi}) L - A(\overline{L} + \overline{\psi} L) + B(L + \psi \overline{L}).$$

$A(\overline{L} + \overline{\psi}L)$ is the $^{\psi}T'$ part. While, by the type of vectors,

$$A\overline{\psi} + B = \xi\overline{\psi}$$
$$A + B\psi = 0.$$

As $A = -B\psi = -(\xi\overline{\psi} - A\overline{\psi})\psi$,

$$A \equiv 0 \bmod \{\xi\overline{\psi}\psi, \overline{\psi}\psi\}.$$

So in our case, this trouble-term does not appear in the linear term. □

Let $R(\rho, h)$ denotes the scalar curvature for this CR structure (the contact form $\frac{1}{\rho^2}\theta$ is fixed). Then, in the real 3-dimensional case, the scalar curvature $R(\rho, h)$ is $d\omega(\rho, h)_1{}^1(\overline{L'}, L')$ in mod $\mid h \mid^2$. Furthermore, by the definition of the CR Hamiltonian flow,

$$\psi \simeq \overline{\partial_{T'}}Z_{\sqrt{-1}h}(\overline{L'}) = L'L'h$$

Proposition 5.2. *The principal part of the linear part of $R(\rho, h)$ with respect to h is*

$$-\rho(\overline{L}\overline{L}LL + LL\overline{L}\overline{L})h.$$

For this part, we have the following main theorem.

Theorem 5.3. $\overline{L}\overline{L}LL + LL\overline{L}\overline{L}$ *is subelliptic.*

Proof. It suffices to show:

$$\|(\overline{L}\overline{L}LL + LL\overline{L}\overline{L})h\| \geq c\|LLh\|, c\|\overline{L}\overline{L}h\|, c\|\overline{L}Lh\|^2, c\|L\overline{L}h\|^2, c\|\xi h\|^2,$$

for a real valued compactly supported C^∞ function h. Here c is a positive constant.

We remember $[\overline{L}, L]_\xi = \sqrt{-1}\xi$. For a real valued C^∞ function h, which is supported, compactly, we compute

$$((\overline{L}\overline{L}LL + LL\overline{L}\overline{L})h, h).$$

First, as for $\|LLh\|^2$, $\|\overline{L}\overline{L}h\|^2$, this is obvious. For the other terms, we modify $(\overline{L}\overline{L}LLu, u)$.

$$(\overline{L}\overline{L}LLh, h) = (\overline{L}[\overline{L}, L]Lh, h) + (\overline{L}L\overline{L}Lh, h)$$
$$= (\overline{L}\sqrt{-1}\xi Lh, h) + \|\overline{L}Lh\|^2$$
$$= (\sqrt{-1}\xi\overline{L}Lh, h) + +\|\overline{L}Lh\|^2.$$

For $(LL\overline{L}\overline{L}h, h)$,

$$(LL\overline{L}\overline{L}h, h) = (L\overline{L}L\overline{L}h, h) + (L[L, \overline{L}]\overline{L}h, h)$$

$$= \|L\overline{L}h\|^2 + (-L\sqrt{-1}\xi\overline{L}h, h)$$
$$= \|L\overline{L}h\|^2 + (-\sqrt{-1}\xi L\overline{L}h, h)$$

While

$$(\sqrt{-1}\xi\overline{L}Lh, h) + (-\sqrt{-1}\xi L\overline{L}h, h)$$

is $(\sqrt{-1}\xi[\overline{L}, L]u, u)$. That is to say

$$(\sqrt{-1}\xi(\sqrt{-1}\xi)h, h) = (-\xi^2 h, h)$$
$$= (\xi h, \xi h).$$

So, we have estimated

$$\|\overline{L}Lh\|^2, \|L\overline{L}h\|^2, \|\xi h\|^2.$$

Hence we have our theorem. $\qquad\qquad\square$

This theorem suggests that in our model case (the contact structure is $\frac{1}{\rho^2}\theta$), if ρ is sufficiently close to 1, then there is a CR Hamiltonian flow, generated by a real valued C^∞ function h, satisfying that the corresponding scalar curvature vanishes, that is to say, $R(\rho, h) = 0$. For the proof, we have to see the non-linear part more carefully. But, I expect that in the more general setting, more complete theorem might be obtained, and so, in a future paper, we will discuss this problem.

References

A1.   T. Akahori, *Complex analytic construction of the Kuranishi family on a normal strongly pseudoconvex manifold*, Publ. Res. Inst. Math. Sci. **14** (1978), 789-847.

A2.   T. Akahori, *The new Neumann operator associated with deformations of strongly pseudo convex domains and its application to deformation theory*, Inventiones Mathematicae **68** (1982). 317-352,

A3.   T. Akahori, *The new estimate for the subbundles E_j and its application to the deformation of the boundaries of strongly pseudo convex domains*, Inventiones Mathematicae **63** (1981), 311–334.

A4.   T. Akahori, *The notion of CR Hamiltonian flows and the local embedding problem of CR structures*, Nova Publisher, 79–94, ISBN 978-1-60741-011-9.

AGL1.   T. Akahori, P. M. Garfield, and J. M. Lee, *Deformation theory of five-dimensional CR structures and the Rumin complex*, Michigan Mathematical Journal, **50** (2002), 517–549.

Mi.   K. Miyajima, *CR construction of the flat deformations of normal isolated singularities*, J. Algebraic Geometry, **8** (1999), 403-470.

12

Ku.   M. Kuranishi, *Application of $\overline{\partial}_b$ to deformations of isolated singularities*, Proc. Sympos. Pure Math., Vol. XXX, Part 1, Williams Coll., Williamstown, Mass., 1975, 97–106, Amer. Math. Soc. , Providence, RI. 1977.

Ku1.   M. Kuranishi, *Local geometry of nondegenerate CR structures* (Mexico, 1983), 1–36, Res. Notes in Math. **112**, Pitman, Boston, MA 1985.

J-L1.   D. Jerison and J. M. Lee, *Intrinsic CR normal coordinates and the CR Ymabe problem*, J. Differential Geometry **29** (1989), 303-343.

An example of the reduction of a single ordinary differential equation to a system, and the restricted Fuchsian relation

Kana Ando

Department of Mathematics, Chiba University, Chiba, 263-8522, Japan
ando@graduate.chiba-u.jp

In the paper [AK], we considered the reduction of a single linear differential equation which has a finite number of regular singular points and an irregular singular point to a system of linear differential equations, where all of the regular singular points are mutually distinct. Furthermore in the paper [A], we considered such a reduction problem, in which regular singular points are not necessarily distinct. In this paper, we shall review our method of reduction, and we shall describe an explicit example of our reduction, we shall compare and compute the restricted Fuchsian relations of the single differential equation and the system of differential equations.

Keywords: regular singular point, irregular singularity, differential equation
AMS classification numbers: 34M25

1. Introduction

In this paper, we shall introduce an example of a reduction from a single differential equation with a finite number of regular singular points and one irregular singular point to a system of linear differential equations whose coefficients are polynomials of degree one. We also compute and compare the corresponding Fuchsian relations, which are an important step in understanding the Stokes phenomenon. In particular, the determination of a Fuchsian relation can be used to solve of the connection problem, which concerns the relationship between the asymptotic expansions of solutions of differential equations in different sectors.

In the paper [AK], we considered the reduction from a linear differential equation which has a finite number of regular singular points and one irregular singular point of rank one at infinity in the complex projective line to a system of linear differential equations whose coefficients are polynomials of degree one. In the paper [A], we considered a more general reduction problem, where we relaxed the conditions on the regular singularities. In both papers, we showed that the system we obtain can be reduced to a generalized Schlesinger system. This result is promising, because it is intimately related with the connection problem. Indeed, in the last section of the paper [K2], Kohno sketched a method for finding the solutions of a multi-point connection problem for a generalized Schlesinger system. In order to solve the connection problem, we need a Fuchsian relation, and such a relation is found in [H]. In general, a Fuchsian relation is given by the sum of the characteristic exponents of all of the solutions for differential equations, but for the purposes of solving the connection problem, we are interested only in the exponents for nonholomorphic solutions. Therefore, we shall call this restricted sum the restricted Fuchsian relation.

In the second section, we shall review the reduction that we proved in [AK] and [A]. In the third section, we shall describe an example of our reduction from a single differential equation to a generalized Schlesinger system. We believe that an explicit example is valuable for understanding our algorithm of the reduction. Furthermore, it has been helpful in our work on the multi-point connection problem, which should appear in the future. In the fourth section, we shall give an algebraic equation for determinating the characteristic exponents for the regular singular points of both a single differential equation and a system of differential equations. We find that the difference between the sum of the characteristic exponents of all nonholomorphic solutions at regular singular points for a single differential equation and a system of differential equations is always an integer. For that purpose, it is enough to know the algebraic equation which the characteristic exponents satisfy because of the relation of solutions and coefficients. This fact means that our reduction preserves monodromic properties of the solutions for differential equations at the regular singular points. We shall show that the nonholomorphic solutions for the single differential equation satisfy the restricted Fuchsian relation, by employing a method of Kohno in the last section. That is, we shall take a sum of the characteristic exponents obtained in the fourth section, and the characteristic exponents for irregular singular points.

We recall the definition of characteristic exponents and constants, and the Fuchisan relation. For the rest of this paper, we assume that t is a complex variable. Consider an n-th order single linear differential equation, with unknown function y, of the form:

$$P_n(t)\frac{d^n y}{dt^n} = P_{n-1}(t)\frac{d^{n-1}y}{dt^{n-1}} + \cdots + P_1(t)\frac{dy}{dt} + P_0(t)y, \qquad (1.1)$$

with a finite number of regular singular points at $t = \lambda_\nu$ ($\nu = 1, 2, \ldots, q$ for some $1 \leq q \leq n$) and one irregular singular point of rank one at $t = \infty$ in the complex projective line, such that each coefficient $P_j(t)$ ($j = 0, 1, \ldots, n - 1, n$) is a complex polynomial of degree at most n. It is well-known that under these assumptions the roots of $P_n(t)$ are the regular singular points. Furthermore, in a punctured disc $0 < |t - \lambda_\nu| < r := \min\{|\lambda_\nu - \lambda_i| : i \neq \nu, i = 1, 2, \ldots, q\}$ ($\nu = 1, 2, \ldots, q$), there exists at least one solution of (1.1) of the form

$$y(t) = (t - \lambda_\nu)^\rho \sum_{m=0}^{\infty} g(m) (t - \lambda_\nu)^m, \quad g(0) \neq 0, \tag{1.2}$$

and we can easily see that the $t = \lambda_\nu$ ($\nu = 1, 2, \ldots, q$) are regular singular points of the fundamental solutions. We call ρ the characteristic exponent of the solution y. If ρ is a positive integer, the solution is holomorphic at λ_ν and if not, the solution is non-holomorphic at λ_ν. At the irregular singular point $t = \infty$, we can find formal solutions of (1.1) of the form

$$y(t) = e^{\mu t} t^\eta \sum_{s=0}^{\infty} h(s) t^{-s}, \quad h(0) \neq 0. \tag{1.3}$$

We call η the characteristic exponent and μ the characteristic constant of this formal solution. We are interested in the sum of characteristic exponents over all non-holomorphic solutions.

2. Method of the reduction

We review the method of our reduction of the papers [AK] and [A]. We recall the differential equation (1.1):

$$P_n(t) \frac{d^n y}{dt^n} = P_{n-1}(t) \frac{d^{n-1} y}{dt^{n-1}} + \cdots + P_1(t) \frac{dy}{dt} + P_0(t) y.$$

We now assume that

$$P_n(t) = (t - \lambda_1)^{n_1} (t - \lambda_2)^{n_2} \cdots (t - \lambda_q)^{n_q} \tag{2.4}$$

with $n_1 + n_2 + \cdots + n_q = n$ ($1 \leq n_q \leq n_{q-1} \leq \cdots \leq n_1 \leq n$).

In order for $t = \lambda_\nu$ ($\nu = 1, 2, \ldots, q$) to be a regular singularity of (1.1), it is necessary and sufficient for the functions

$$\frac{(t - \lambda_\nu)^i P_{n-i}(t)}{P_n(t)} \quad (i = 1, 2, \ldots, n)$$

to be holomorphic at $t = \lambda_\nu$. Hence, for each ν, the polynomials $P_{n-i}(t)$ ($1 \leq i \leq n_\nu$) have the factor $(t - \lambda_\nu)^{n_\nu - i}$. It is therefore easy to see that the coefficients $P_{n-i}(t)$ are written as:

$$
\begin{cases}
P_{n-i}(t) = \left[\displaystyle\prod_{\nu=1}^{q}(t-\lambda_\nu)^{n_\nu-i}\right]\widehat{P}_{n-i}(t) & (0 < i \le n_q), \\[3mm]
P_{n-i}(t) = \left[\displaystyle\prod_{\nu=1}^{k}(t-\lambda_\nu)^{n_\nu-i}\right]\widehat{P}_{n-i}(t) & (n_{k+1} < i \le n_k\,;\, k = 1, 2, \ldots, q), \\[3mm]
P_{n-i}(t) = \widehat{P}_{n-i}(t) & (n_1 < i \le n)
\end{cases}
$$

$$(2.5)$$

where $\widehat{P}_{n-i}(t)$ is a polynomial for all $i = 0, 1, \ldots, n$ and n_{q+1} is equal to 1. The notation $\widehat{P}_{n-i}(t)$ will appear again in the fourth section. In the paper [A], we considered the reduction of the single differential equation (1.1) to the system of differential equations, with unknown length n vector function Y

$$(tI - B)\frac{dY}{dt} = (A + C\,t)\,Y, \qquad (2.6)$$

where I is the n by n identity matrix, A is an n by n constant matrix, C is an n by n constant lower triangular matrix, and

$$
B = \operatorname{diag}(\overbrace{\lambda_1, \cdots, \lambda_1}^{n_1},\ \overbrace{\lambda_2, \cdots, \lambda_2}^{n_2},\ \cdots,\ \overbrace{\lambda_q, \cdots, \lambda_q}^{n_q}).
$$

In order to reduce (1.1) to (2.6), we apply the transformation:

$$
\begin{cases}
y_1 = y, \\
y_2 = \varphi_1\, y' + e_{2,0}(t)\, y, \\
\quad \vdots \\
y_j = \varphi_{j-1}\, y^{(j-1)} + e_{j,j-2}(t)\, y^{(j-2)} + \cdots + e_{j,1}(t)\, y' + e_{j,0}(t)\, y, \\
\quad \vdots \\
y_n = \varphi_{n-1}\, y^{(n-1)} + e_{n,n-2}(t)\, y^{(n-2)} + \cdots + e_{n,1}(t)\, y' + e_{n,0}(t)\, y,
\end{cases}
$$

$$(2.7)$$

where we use the following notation:

- We define $\{y_j\}_{j=1}^{n}$ by using the unknown function y.
- We define φ_j by the following:

$$
\varphi_j = \prod_{k=1}^{j}(t - \lambda_{\sigma(k)}), \qquad (j = 1, 2, \ldots, n),
$$

where $\sigma : \{1, 2, \ldots, n\} \to \{1, 2, \ldots, q\}$ is the unique weakly increasing function such that $\sigma^{-1}(i)$ has cardinality n_i for all $1 \le i \le q$.

- $e_{j,\,j-i}(t)(j = 2, \ldots, n, i = 1, 2, \ldots, j)$ is a polynomial in t.
- y' and $y^{(j)}(j = 1, 2, \ldots, n)$ mean the first and jth derivatives of y with respect to t, respectively.

By using (2.7), we can derive the system of linear differential equations for the column vector $Y = (y_1, y_2, \ldots .y_n)^t$:

$$\begin{pmatrix} t - \lambda_{\sigma(1)} & & & \text{\Large 0} \\ & t - \lambda_{\sigma(2)} & & \\ & & \ddots & \\ & & & \ddots \\ \text{\Large 0} & & & t - \lambda_{\sigma(n)} \end{pmatrix} \frac{dY}{dt} = \begin{pmatrix} d_{1,1} & 1 & & & \text{\Large 0} \\ d_{2,1} & d_{2,\,2} & 1 & & \\ \vdots & \vdots & \ddots & \ddots & \\ \vdots & \vdots & & \ddots & 1 \\ d_{n,1} & d_{n,2} & \cdots & \cdots & d_{n,n} \end{pmatrix} Y,$$

where all $d_{j,i} = d_{j,\,i}(t)$ are degree one polynomials, and $Y = (y_1, y_2, \ldots, y_n)^t$ means the transpose of the indicated row vector. This system is our candidate for the reduction from (1.1) to (2.6). In other words, we define $d_{j,i}$ by reducing (1.1) using (2.7). Details will appear in the paper [A].

Theorem 2.1. *We define the entries $d_{j,i}(t)(j = 1, 2, \ldots, n, i = 1, 2, \ldots, j)$ of the matrix $(A + Ct)$ in (2.6) by applying the transformation (2.7) to (1.1). Then we obtain the following relation:*

$$(t - \lambda_j)\left(e_{j,\,j-\ell-3}(t) + e'_{j,\,j-\ell-2}(t)\right)$$
$$= e_{j+1,\,j-\ell-2}(t) + \sum_{h=0}^{\ell+1} d_{j,\,j-h}(t)\, e_{j-h,\,j-\ell-2}(t) \quad (2.8)$$
$$(j = 1, 2, \ldots, n\,;\, \ell = -1, 0, \ldots, j-2)$$

where

- $e_{n+1,\,k}(t) = -P_k(t)$ *for* $k = 0, 1, \ldots, n-1$,
- $e_{j,\,-1}(t) \equiv 0$,
- $e'_{j,\,j-\ell-2}(t)$ *means the first derivative with respect to t,*
- *all $d_{j,\,i}(t)$ are polynomials of degree one.*

We will not describe the proof in detail, but we remark that it follows from comparing coefficients of t in (2.8). See [AK] and [A] for details.

We showed that after multiplying by $(tI - B)^{-1}$ on the left side of (2.6), the equation is reduced to the following generalized Schlesinger system:

$$\frac{dY}{dt} = \left(\sum_{i=1}^{q} \frac{\bar{A}_i}{t - \lambda_i} + C\right) Y$$

where $\bar{A}_i (i = 1, 2, \ldots, q)$ are n by n constant matrices. In future work, we intend to use this result to find the solutions of a multi-point connection problem (see [K2]).

3. Example of the reduction

In this section, we consider an example of the reduction of a fourth order linear differential equation. We apply our theory, with unknown function y to the following equation

$$t (t - 1)^3 y^{(4)} = P_3(t) y^{(3)} + P_2(t) y'' + P_1(t) y' + P_0(t) y, \qquad (3.9)$$

where

$$\begin{cases} P_3(t) &= -(t-1)^2 (5t+1), \\ P_2(t) &= -(t-1)^2 (t-4), \\ P_1(t) &= t^4 - t^3 - t^2 + 21t - 8, \\ P_0(t) &= 4t^3 - 3t^2 + 4t + 3. \end{cases}$$

Obviously, this linear differential equation has regular singularities at $t = 0, 1$ and an irregular singular point of rank one at infinity. Now, we consider the reduction of (3.9) to a system of linear differential equations of the form $(tI - B) Y' = (A + Ct) Y$ by the transformation

$$\begin{cases} y_1 &= y, \\ y_2 &= \varphi_1 y' + e_{2,0}(t) y, \\ y_3 &= \varphi_2 y'' + e_{3,1}(t) y' + e_{3,0}(t) y, \\ y_4 &= \varphi_3 y^{(3)} + e_{4,2}(t) y'' + e_{4,1}(t) y' + e_{4,0}(t) y, \end{cases}$$

where I is the 4×4 identity matrix, A and C are 4 by 4 constant matrices with C lower triangular, B is a diagonal matrix:

$$B = \mathrm{diag}(1, 1, 1, 0),$$

$\{y_j\}_{j=1}^4$ are defined using the unknown function y from (3.9), $\varphi_1 = t - 1$, $\varphi_2 = (t - 1)^2$, $\varphi_3 = (t - 1)^3$, and the coefficients $e_{i,j}(t)$ ($i = 2, 3, 4, j = 0, \ldots, i - 2$) are polynomials in t.

Then, we have the following relations:

① $t(e_{4,2}(t) + \varphi_3') = -P_3 + d_{4,4} \varphi_3,$

② $t(e_{4,1}(t) + e_{4,2}'(t)) = -P_2 + d_{4,4} e_{4,2}(t) + d_{4,3} \varphi_2,$

③ $t(e_{4,0}(t) + e_{4,1}'(t)) = -P_1 + d_{4,4} e_{4,1}(t) + d_{4,3} e_{3,1}(t) + d_{4,2} \varphi_1,$

④ $t e_{4,0}'(t) = -P_0 + d_{4,4} e_{4,0}(t) + d_{4,3} e_{3,0}(t) + d_{4,2} e_{2,0}(t) + d_{4,1},$

⑤ $(t-1)(e_{3,1}(t) + \varphi_2') = e_{4,2}(t) + d_{3,3}\,\varphi_2,$

⑥ $(t-1)(e_{3,0}(t) + e_{3,1}'(t)) = e_{4,1}(t) + d_{3,3}\,e_{3,1}(t) + d_{3,2}\,\varphi_1,$

⑦ $(t-1)e_{3,0}'(t) = e_{4,0}(t) + d_{3,3}\,e_{3,0}(t) + d_{3,2}\,e_{2,0}(t) + d_{3,1},$

⑧ $(t-1)(e_{2,0}(t) + \varphi_1') = e_{3,1}(t) + d_{2,2}\,\varphi_1,$

⑨ $(t-1)e_{2,0}'(t) = e_{3,0}(t) + d_{2,2}\,e_{2,0}(t) + d_{2,1},$

⑩ $-e_{2,0}(t) = d_{1,1}.$

In the above, $d_{j,i}$ are polynomials of degree 1 in t. Avoiding the difficult calculation, we shall show just the order of steps in our reduction algorithm. First, we calculate the principal diagonal elements $d_{i,i}(t)$ ($i = 4, 3, 2, 1$). We follow the order of calculation ① → ⑤ → ⑧ → ⑩.

Next, we shall proceed to the calculation of the first subdiagonal elements $d_{i,i-1}(t)$ ($i = 4, 3, 2$) by following ② → ⑥ → ⑨.

In order to determine the second subdiagonal elements $d_{i,i-2}(t)$ ($i = 4, 3$), we follow the order of calculations ③ → ⑦.

Lastly, from ④ we obtain the value of $d_{4,1}(t)$.

We have thus determined all coefficients $e_{i,j}(t)$ of the transformation and the elements of $d_{j,i}(t)$ as follows:

$$
\begin{cases}
e_{2,0}(t) &= -\omega^2\,(t-1), \\
e_{3,0}(t) &= (t-1)(t-\omega), \\
e_{3,1}(t) &= (t-1)^2, \\
e_{4,0}(t) &= -t^3 + 3t^2 + (\omega - 3)t - (\omega + 5), \\
e_{4,1}(t) &= 3(t-1)(t-3), \\
e_{4,2}(t) &= 3(t-1)^2,
\end{cases}
$$

$$
\begin{cases}
d_{1,1} &= \omega^2\,(t-1), \\
d_{2,1} &= (\omega - 1)(t-1), \\
d_{2,2} &= \omega\,(t-1) + 1, \\
d_{3,1} &= -(9\omega + 4)(t-1) + 6, \\
d_{3,2} &= -(\omega - 1)(t-1) + 6, \\
d_{3,3} &= (t-1) - 1, \\
d_{4,1} &= -(17\omega + 12)t + (18\omega + 26), \\
d_{4,2} &= (\omega + 2)t + 18, \\
d_{4,3} &= 2t + 1, \\
d_{4,4} &\; -1,
\end{cases}
$$

where ω is a non-real root of $\omega^3 - 1 = 0$.

Consequently, we can reduce the single linear differential equation (3.9) to a system of linear differential equations of the form

$$(t\,I - B)\frac{dY}{dt} = (A + C\,t)\,Y = \{\text{diag}\,((t-1)I_{n_1},\,t)\,C + \bar{A}\}Y, \quad (3.10)$$

where for the rest of this paper, I_ν denotes the ν by ν identity matrix, and A, $\bar{A}$ and C are the constant matrices given as follows:

$$A = \begin{pmatrix} -\omega^2 & 1 & 0 & 0 \\ -(\omega - 1) & -(\omega - 1) & 1 & 0 \\ 9\omega + 10 & \omega + 5 & -2 & 1 \\ 18\omega + 26 & 18 & 1 & 1 \end{pmatrix},$$

$$\bar{A} = \begin{pmatrix} \bar{a}_1 & 1 & 0 & 0 \\ \bar{a}_{2,1} & \bar{a}_2 & 1 & 0 \\ \bar{a}_{3,1} & \bar{a}_{3,2} & \bar{a}_3 & 1 \\ \bar{a}_{4,1} & \bar{a}_{4,3} & \bar{a}_{4,3} & \bar{a}_4 \end{pmatrix} = \begin{pmatrix} 0 & 1 & 0 & 0 \\ 0 & 1 & 1 & 0 \\ 6 & 6 & -1 & 1 \\ 18\omega + 26 & 18 & 1 & 1 \end{pmatrix},$$

and

$$C = \begin{pmatrix} \omega^2 & 0 & 0 & 0 \\ \omega - 1 & \omega & 0 & 0 \\ -(9\omega + 4) & -(\omega - 1) & 1 & 0 \\ -17\omega - 12 & \omega + 2 & 2 & 0 \end{pmatrix}.$$

We will describe how to compute the entries of $\bar{A}$ in the next section.

4. Characteristic exponents and constants

We shall investigate the characteristic exponents for the regular singular points and the characteristic constants for the irregular singular point for the differential equation (1.1) and the output (2.6) of the reduction. In order to show that our reduction preserves monodromic properties of the solutions for differential equations at the regular singular points, we will show that the difference between the sum of the characteristic exponents for the regular singular points of (1.1) and the corresponding sum for (2.6) is an integer. The calculation of this difference will appear in the proposition at the end of this section. In the reduction from (1.1) to (2.6), all entries $e_{j,i}(t)$ of the transformation matrix are polynomials in t. We shall show here that the characteristic exponents at each regular singular point of both (1.1) and (2.6) are invariant modulo integers.

We recall the notation of [A] for obtaining the restricted Fuchsian relation of (1.1) and (2.6). We introduce the notation N_k, f_k^i and ψ_k where $k = 1, 2, \cdots, q$ and i is an integer index (not an exponent).

$$
\begin{cases}
N_k = n_1 + n_2 + \cdots + n_k \quad (k = 1, 2, \ldots, q), \\
f_k^i = \displaystyle\prod_{\nu=1}^{k} (t - \lambda_\nu)^{n_\nu - i}, \\
\psi_k = \displaystyle\prod_{\nu=1}^{k} (t - \lambda_\nu),
\end{cases}
$$

where $N_0 \equiv 0$, then we know that

$$
\begin{cases}
N_q = n, \\
f_k^i = f_k^{i+1} \psi_k.
\end{cases}
$$

Then, we can rewrite the coefficients (2.5) of (1.1) as follows:

$$
\begin{cases}
P_{n-i}(t) = f_k^i \, \widehat{P}_{n-i}(t) \ (n_{k+1} < i \leq n_k \,;\, k = 1, 2, \ldots, q), \\
P_{n-i}(t) = \widehat{P}_{n-i}(t) \quad (n_1 < i \leq n)
\end{cases}
$$

where $n_{q+1} \equiv 0$.

As we saw in the introduction, in the punctured disc $0 < |t - \lambda_\nu| < r(\nu = 1, 2, \ldots, q)$, there exists at least one solution of (1.1)

$$
y(t) = (t - \lambda_\nu)^\rho \sum_{m=0}^{\infty} g(m) \, (t - \lambda_\nu)^m.
$$

Substituting it into (1.1), we find that the characteristic exponent ρ is a root of the equation

$$
[\rho]_n = \sum_{i=1}^{n_\nu} \gamma_i \, [\rho]_{n-i},
$$

where $[\rho]_k$ is the Pochhammer symbol for $k = 0, 1, 2, \ldots$, defined by the following recursion:

$$[\rho]_k = \rho(\rho - 1) \cdots (\rho - k + 1), \quad [\rho]_0 \equiv 1,$$

and the coefficients γ_i are given by

$$\gamma_i = \left[\frac{P_{n-i}(t)}{P_n(t)} (t - \lambda_\nu)^i \right]_{t=\lambda_\nu}$$

$$= \frac{\widehat{P}_{n-i}(\lambda_\nu)}{\displaystyle\prod_{\substack{\ell=1 \\ \ell \neq \nu}}^{k-1} (\lambda_\nu - \lambda_\ell)^i \prod_{\ell=k}^{q} (\lambda_\nu - \lambda_\ell)^{n_\ell}} \quad (n_k < i \leq n_{k-1} \leq n_\nu).$$

Then ρ is a root of

$$[\rho]_{n-n_\nu} = 0, \tag{4.11}$$

or a root of

$$[\rho - n + n_\nu]_{n_\nu} = \sum_{i=1}^{n_\nu} \gamma_i \, [\rho - n + n_\nu]_{n_\nu - i}. \tag{4.12}$$

For (1.1), (4.11) and (4.12) imply that there exist $n - n_\nu$ holomorphic solutions and n_ν possibly nonholomorphic solutions in the punctured disc $0 < |t - \lambda_\nu| < r$.

Next, we shall consider the characteristic exponents and constants for (2.6). For (2.6), there also exist $(n - n_\nu)(\nu = 1, 2, \ldots, q)$ holomorphic solutions and n_ν possibly nonholomorphic solutions near each singular point $t = \lambda_\nu$. For the calculation, we shall rewrite (2.6) by setting $\bar{a}_j := a_j - \lambda_k c_j$, and $\bar{a}_{j,i} := a_{j,i} - \lambda_k c_{j,i}$ where a_j and $a_{j,i}$ are entries of A and c_j and $c_{j,i}$ are entries of C of (2.6). $a_{j,i}$ and $c_{j,i}$ are the (j,i)-entries of A and C, respectively, and $a_j := a_{j,j}$ and $c_j := c_{j,j}$ are j-th diagonal entries. That is, we obtain the formula:

$$\begin{cases} d_{j,j}(t) = c_j t + a_j = c_j (t - \lambda_k) + \bar{a}_j, \\ d_{j,i}(t) = c_{j,i} t + a_{j,i} = c_{j,i} (t - \lambda_k) + \bar{a}_{j,i} \\ (N_{k-1} < j \leq N_k ; \, k = q, q-1, \ldots, 1, i = 1, 2, \ldots, j-1). \end{cases}$$

Then we rewrite the right hand side of (2.6) in the form

$$\{\operatorname{diag}((t - \lambda_1) I_{n_1}, (t - \lambda_2) I_{n_2}, \cdots, (t - \lambda_q) I_{n_q}) C + \bar{A}\} Y,$$

where C is a lower triangular constant matrix, and $\bar{A}$ is of the form

$$\bar{A} = \begin{pmatrix} \bar{a}_1 & 1 & & & \\ \bar{a}_{2,1} & \bar{a}_2 & 1 & & \text{\Large 0} \\ \bar{a}_{3,1} & \bar{a}_{3,2} & \ddots & \ddots & \\ \vdots & \vdots & \ddots & \ddots & 1 \\ \bar{a}_{n,1} & \bar{a}_{n,2} & \cdots & \bar{a}_{n,n-1} & \bar{a}_n \end{pmatrix}.$$

In [A], for finding the order of $e_{j,\,j-i}(t)$, we showed that for $k = 1, 2, \ldots, q$

$$e_{j,\,j-i}(t)$$
$$= f_k^{i-1}(t - \lambda_k)^{j-N_k}\,\widehat{e}_{j,\,j-i}(t) \quad (N_{k-1} < j \leq N_k, i = 1, 2, \ldots, j-1)$$

where $(t - \lambda_k)^p \equiv 1$ $(p \leq 0)$ and $\widehat{e}_{j,\,j-i}(t)$ is a polynomial of t, and that for $1 \leq i \leq n_\nu$ we obtained the explicit form of $\bar{A}$ as follows:

$$\bar{A}_\nu = \begin{pmatrix} 0 & 1 & & & & \\ & 1 & 1 & & & \\ & & & \ddots & & \ddots \\ \text{\Large 0} & & & & n_\nu - 2 & 1 \\ \bar{a}_{N_\nu,\,N_\nu-1+1} & \bar{a}_{N_\nu,\,N_\nu-1+2} & \cdots\cdots & \bar{a}_{N_\nu,\,N_\nu-1} & \bar{a}_{N_\nu} \end{pmatrix},$$

where $\bar{A}_\nu$ is the ν by ν block-diagonal matrix of $\bar{A}$, and

$$\bar{a}_{N_\nu,\,N_\nu-i+1} = -\frac{\widehat{e}_{N_\nu+1,\,N_\nu-i}(\lambda_\nu)}{(\psi_{\nu-1}(\lambda_\nu))^i}, \tag{4.13}$$

$$\bar{a}_{N_\nu} = n_\nu - 1 + \bar{a}_{N_\nu,N_\nu}.$$

The characteristic exponents of nonholomorphic solutions of (2.6) are given by eigenvalues of the constant matrix $\bar{A}_\nu$. Since the matrix $\bar{A}_\nu$ is a companion matrix, the eigenvalues are roots of the equation

$$[\widehat{\rho}]_{n_\nu} = \sum_{i=1}^{n_\nu} \bar{a}_{N_\nu,\,N_\nu-i+1}[\widehat{\rho}]_{n_\nu-i}. \tag{4.14}$$

Now we shall show that the reduction described above preserves characteristic properties.

Proposition 4.1. *The sum of the characteristic exponents of nonholomorphic solutions of (2.6) differ from the sum of the characteristic exponents of nonholomorphic solutions of (1.1) at each regular singular point $t = \lambda_\nu$ only by the integers $n_\nu N_{\nu-1}$.*

Proof. We define

$$\rho' := \widehat{\rho} - n + N_\nu.$$

Substituting

$$\rho' + n - N_\nu$$

into $\widehat{\rho}$ of (4.14), we find that from (4.13), for $\nu = 1, 2, \ldots, q$

$$[\rho']_{n_\nu} = \sum_{i=1}^{n_\nu} \gamma_i \, [\rho']_{n_\nu - i}, \tag{4.15}$$

but the proof is an induction argument that we omit. By the relation between the coefficients of a polynomial and its roots for (4.15), (4.12), and (4.14) we have thus verified that:

$$\sum_{\ell=1}^{n_\nu} \rho_{\nu,\ell} - \sum_{\ell=1}^{n_\nu} \rho'_{\nu,\ell} = n_\nu(n - n_\nu)$$

and

$$\sum_{\ell=1}^{n_\nu} \rho'_{\nu,\ell} - \sum_{\ell=1}^{n_\nu} \widehat{\rho}_{\nu,\ell-} = n_\nu(-n + N_\nu).$$

By adding the two formulas above, we obtain the desired equation:

$$\sum_{\ell=1}^{n_\nu} \rho_{\nu,\ell} - \sum_{\ell=1}^{n_q} \widehat{\rho}_{\nu,\ell} = n_\nu N_{\nu-1}. \qquad \square$$

This proposition means that the transformation treated between (1.1) and (2.6) in this paper preserves the monodromic properties. We remark that there is a fundamental set of solutions to (2.6) with the form (1.2) in the punctured disk $0 < |t - \lambda_\nu| < r := \min\{|\lambda_\nu - \lambda_i| : i \neq \nu, i = 1, 2, \ldots, q\}$ ($\nu = 1, 2, \ldots, q$) where $g(m)$ is a nonzero n-entry column vector. There also exists a formal solution of (2.6) at the irregular singular point with the form (1.3) where $h(s)$ is a nonzero n-entry column vector. Like the case of a single differential equation, we call the numbers ρ, η characteristic exponents and μ characteristic constants.

5. Restricted Fuchsian relation

Lastly, we shall explain an important identity, which necessarily exists among characteristic exponents for (1.1) and plays an essential role in the global analysis of linear differential equations with regular or irregular singularities.

As in the previous section, in the punctured disc $0 < |t - \lambda_\nu| < r$ ($\nu = 1, 2, \ldots, q$), there exist n_ν non-holomorphic solutions of (1.1)

$$y_{\nu, \ell}(t) = (t - \lambda_\nu)^{\rho_{\nu, \ell}} \sum_{m=0}^{\infty} g_{\nu, \ell}(m) (t - \lambda_\nu)^m \qquad (\ell = 1, 2, \ldots, n_\nu),$$

where the characteristic exponent $\rho_{\nu, \ell}$ are roots of the characteristic equation

$$[\rho - n + n_\nu]_{n_\nu} = \sum_{i=1}^{n_\nu} \gamma_{\nu, i} [\rho - n + n_\nu]_{n_\nu - i}. \tag{5.16}$$

The coefficients $\gamma_{\nu, i}$ are given by

$$\gamma_{\nu, i} = \left[\frac{P_{n-i}(t)}{P_n(t)} (t - \lambda_\nu)^i \right]_{t=\lambda_\nu}. \tag{5.17}$$

We now consider the sum of all characteristic exponents $\rho_{\nu, \ell}$. From the characteristic equation (5.16), we immediately obtain

$$\sum_{\ell=1}^{n_\nu} \rho_{\nu, \ell} = \sum_{\ell=1}^{n_\nu} (n - \ell) + \gamma_{\nu, 1}$$

$$= n \, n_\nu - \frac{n_\nu(n_\nu + 1)}{2} + \gamma_{\nu, 1}.$$

Then we have

$$\sum_{\nu=1}^{q} \sum_{\ell=1}^{n_\nu} \rho_{\nu, \ell} = \left(n - \frac{1}{2} \right) \sum_{\nu=1}^{q} n_\nu - \frac{1}{2} \sum_{\nu=1}^{q} n_\nu^2 + \sum_{\nu=1}^{q} \gamma_{\nu, 1}$$

$$= \left(n - \frac{1}{2} \right) n - \frac{1}{2} \sum_{\nu=1}^{q} n_\nu^2 + \sum_{\nu=1}^{q} \gamma_{\nu, 1}. \tag{5.18}$$

In order to calculate the last sum in the above formula, we apply the fact

$$\gamma_{\nu, 1} = \left[\frac{P_{n-1}(t)}{P_n(t)} (t - \lambda_\nu) \right]_{t=\lambda_\nu}.$$

The right hand side is the residue of $\frac{P_{n-1}(t)}{P_n(t)} dt$ at $t = \lambda_\nu$. We can then express the sum of $\gamma_{\nu,1}$ in the form

$$\sum_{\nu=1}^{q} \gamma_{\nu,1} = \frac{1}{2\pi i} \int_{|t|=R} \frac{P_{n-1}(t)}{P_n(t)} \, dt, \qquad (5.19)$$

where all $t = \lambda_\nu$ ($\nu = 1, 2, \ldots, q$) are included in the disk $|t| < R$ for sufficiently large R, and the path of integration is oriented counterclockwise. According to the theory of residues for rational functions, the right hand side of (5.19) is equal to minus one times the value of the residue at infinity. If we denote the polynomial $P_{n-\ell}(t)$ of degree n by

$$P_{n-\ell}(t) = \sum_{j=0}^{n} p_{n-\ell,j} \, t^j \qquad (\ell = 0, 1, \ldots, n),$$

then for sufficiently large values of t the integrand can be written as follows:

$$\frac{P_{n-1}(t)}{P_n(t)} = \frac{p_{n-1,n} + p_{n-1,n-1}t^{-1} + \cdots + p_{n-1,0}\,t^{-n}}{1 + p_{n,n-1}t^{-1} + \cdots + p_{n,0}\,t^{-n}}$$

$$= (p_{n-1,n} + p_{n-1,n-1}t^{-1} + \cdots + p_{n-1,0}t^{-n})(1 - p_{n,n-1}t^{-1} + \cdots)$$

$$= p_{n-1,n} + (p_{n-1,n-1} - p_{n-1,n}\, p_{n,n-1})t^{-1} + \cdots .$$

Consequently, we have

$$\sum_{\nu=1}^{q} \gamma_{\nu,1} = p_{n-1,n-1} - p_{n-1,n}\, p_{n,n-1}. \qquad (5.20)$$

Now we shall investigate the characteristic exponents of formal solutions for (1.1) at the irregular singularity $t = \infty$, which are expressed in the form

$$y(t) = e^{\mu t} t^{\eta} \sum_{s=0}^{\infty} h(s)\, t^{-s},$$

where we assume that $h(0) \neq 0$, and we define $h(-s) \equiv 0$ when s is a positive integer. We begin with some preparative calculations for finding the characteristic exponent η, following the method in the paper [K1]. We define $y_k(t)(k = 0, 1, \ldots, n)$ to be the kth derivative of $y(t)$ with respect to t:

$$y_k(t) = \frac{d^k y(t)}{dt^k},$$

and we shall write the coefficients of the formal series $h_k(s)$, that is

$$y_k(t) = e^{\mu t} t^{\eta} \sum_{s=0}^{\infty} h_k(s)\, t^{-s}$$

where

$$y_0(t) \equiv y(t).$$

Then, we have the following result:

Lemma 5.1. *From* $y_k(t) = y'_{k-1}(t)(k = 1, 2, \ldots, n)$, *the relation*

$$h_k(s) = \mu\, h_{k-1}(s) + (\eta - s + 1)\, h_{k-1}(s - 1) \quad (s = 0, 1, \ldots) \qquad (5.21)$$

holds, where $h_k(-s) = 0$ *for* $s > 0$.

Proof.

$$y_k(t) = y'_{k-1}(t) \Leftrightarrow e^{\mu t} t^\eta \sum_{s=0}^{\infty} h_k(s)\, t^{-s} = e^{\mu t} t^\eta \left\{ \mu \sum_{s=0}^{\infty} h_{k-1}(s) t^{-s} \right.$$

$$\left. + \sum_{s=0}^{\infty} (\eta - s) h_{k-1}(s) t^{-s-1} \right\}$$

$$\Leftrightarrow \sum_{s=0}^{\infty} h_k(s)\, t^{-s} = \left\{ \mu \sum_{s=0}^{\infty} h_{k-1}(s) t^{-s} \right.$$

$$\left. + \sum_{s=0}^{\infty} (\eta - s) h_{k-1}(s) t^{-s-1} \right\}.$$

Comparing the coefficients of t^{-s}, we have the above formula. $\qquad \square$

Moreover, substituting

$$t^j\, y_k(t) = e^{\mu t} t^\eta \sum_{s=-j}^{\infty} h_k(s + j)\, t^{-s}$$

into (1.1), we have

$$\sum_{j=0}^{n} \left\{ p_{n,j}\, h_n(s + j) - \sum_{k=0}^{n-1} p_{k,j}\, h_k(s + j) \right\} = 0, \quad (s = -j, -j + 1, \ldots).$$

$$(5.22)$$

With this preparation complete, we are now in a position to calculate the value of the characteristic constants μ and the characteristic exponents η of the formal solutions of (1.1). To this end, we iteratively apply (5.21) to get the $k + 1$-term sum (see [K1]):

$$h_k(s) = \mu^k\, h(s) + \{\mu^k + (\eta - s + 1)k\, \mu^{k-1}\} h(s - 1) + \cdots . \qquad (5.23)$$

We then apply (5.22) with the substitution $s = -n$. Then, from (5.23), we obtain

$$\left(\mu^n - \sum_{k=0}^{n-1} p_{k,n}\, \mu^k \right) h(0) = 0.$$

Hence, the equation

$$J(\mu) \equiv \mu^n - \sum_{k=0}^{n-1} p_{k,n}\, \mu^k = 0$$

determines the n characteristic constants, which we shall denote by μ_ℓ ($\ell = 1, 2, \ldots, n$). From here, we assume that they are mutually distinct, i.e., $\mu_\ell \neq \mu_i$ ($\ell \neq i$). Without this assumption, the argument becomes more complicated, because we can no longer use the assumption that $J'(\mu_\ell) \neq 0$ to produce (5.24).

Next, we substitute $s = 1$ into (5.23) to obtain

$$h_k(1) = \mu^k\, h(1) + \{\mu^k + \eta\, k\, \mu^{k-1}\}\, h(0).$$

Then combining this with what we get from substituting $s = -n+1$ into (5.22), we obtain

$$J(\mu)\, h(1) + \{J(\mu) + \eta\, J'(\mu)\}\, h(0)$$

$$+ \left\{ p_{n,n-1}\, \mu^n - \sum_{k=0}^{n-1} p_{k,n-1}\, \mu^k \right\} h(0) = 0,$$

whence the characteristic exponent corresponding to μ_ℓ is given by the formula

$$\eta_\ell = -\, \frac{p_{n,n-1}\, \mu_\ell^n - \sum_{k=0}^{n-1} p_{k,n-1}\, \mu_\ell^k}{J'(\mu_\ell)}. \tag{5.24}$$

By exactly the same consideration as in the case (5.19), we can express the sum of the characteristic exponents (5.24) in the form of the integral

$$\sum_{i=1}^{n} \eta_i = -\, \frac{1}{2\pi i} \int_{|\mu|=R} \frac{p_{n,n-1}\, \mu^n - \sum_{k=0}^{n-1} p_{k,n-1}\, \mu^k}{J(\mu)}\, d\mu$$

for sufficiently large R, with the path of integration oriented counterclockwise. From the residue theorem we obtain

$$\sum_{i=1}^{n} \eta_i = p_{n-1,n-1} - p_{n-1,n}\, p_{n,n-1}. \tag{5.25}$$

Combining this formula with (5.20) and (5.18), we consequently obtain the restricted Fuchs relation.

Theorem 5.2. *Consider (1.1), and set $\rho_{\nu,\ell}$ ($\nu = 1, 2, \ldots, q, \ell = 1, 2, \ldots, n_\nu$) to be the characteristic exponents at regular singular points $t = \lambda_\nu$ and η_i ($i = 1, 2, \ldots, n$) to be the characteristic exponents at the irregular singular point at infinity. Assume the characteristic constants of the formal solutions at the irregular singular point are mutually distinct, i.e., $\mu_\ell \neq \mu_i$ ($\ell \neq i$). Then the restricted Fuchs relation for non-holomorphic solutions for (1.1) is the following:*

$$\sum_{\nu=1}^{q} \sum_{\ell=1}^{n_\nu} \rho_{\nu,\ell} - \sum_{i=1}^{n} \eta_i = \left(n - \frac{1}{2} \right) n - \frac{1}{2} \sum_{\nu=1}^{q} n_\nu^2. \qquad (5.26)$$

Remark 5.3. For the special case $q = 1$, we have $n_1 = n$ and hence the right hand side of (5.26) is equal to $\frac{n(n-1)}{2}$. In particular, the above restricted Fuchs relation is a generalization of the lemma 3.1 in [K1].

References

A.   K. Ando, *On the reduction of a single differential equation to a system of first degree differential equations*, (submitted).

AK. K. Ando, M. Kohno, *A certain reduction of a single differential equation to a system of differential equations*, Kumamoto Journal of Mathematics **19** (2006), 99–114.

H.   M. Hukuhara, *Sur la relation de Fuchs relative a l'equation differentielle lineaire*, Proc. Japan Acad., **34** (1958), 102-106.

K1. M. Kohno, *A Two Point Connection Problem for General Linear Ordinary Differential Equations*, Hiroshima Mathematical Journal **4**, No.2, (1974), 293-338.

K2. M. Kohno, *Global Analysis in Linear Differential Equations*, (Mathematics and Its Applications (Kluwer)) Vol.**471** (1999).

Fronts of weighted cones

Toshizumi Fukui and Masaru Hasegawa

Department of Mathematics, Faculty of Science, Saitama University,
Saitama 338-8570, Japan.
`tfukui@rimath.saitama-u.ac.jp`

Departamento de Matemática, Instituto de Ciências Matemáticas e de Computação,
Universidade de São Paulo - Campus de São Carlos,
Caixa Postal 668, 13560-970 São Carlos, SP, Brazil
`mhasegawa@icmc.usp.br`

We present an asymptotic formula for the principal curvatures and principal directions of weighted cones near the vertex, which is defined by (0.1). As a byproduct, we construct a singular C^3-surface with bounded principal curvatures. We also remark that when the weights are $(1, 2, 2)$ we have focal curves, which is a generalization of focal conics we discussed in [3].

Keywords: singularity of front, weighted cone, principal curvature, principal directions
AMS classification numbers: 53A05, 35A18

Wave fronts (or simply, fronts) is a locus of points in space reached by a wave or vibration at the same instant as the wave travels through a medium. Fronts may have singularities at some moment, and to investigate the singularity types is one of the main topics in the application of singularity theory. The classification of singularity types of generic fronts is known ([1, page 336]), and the local classification of bifurcations in generic one parameter families of fronts in 3-dimensional spaces are also given in [1, page 348].

Fronts of a regular surface in the 3-dimensional Euclidean space is described by parallel surfaces and criteria of its singularity type in terms of differential geometric languages are given in [2]. We also gave ([3]) a similar criteria for fronts of Whitney umbrella, which is only stable singularities type from a plane to the 3-dimensional Euclidean space $\mathbf{R}^3$. Since there are singularities of natural objects in the real world, we are motivated to study singularities of fronts of singular surfaces in $\mathbf{R}^3$. In this paper, we investigate criteria of singularity types of fronts of weighted cones

$$g : \mathbf{R} \times S^1 \to \mathbf{R}^3, \ (r, s) \mapsto (r^{w_1}\gamma_1(s), \ r^{w_2}\gamma_2(s), \ r^{w_3}\gamma_3(s)), \tag{0.1}$$

which has singularities along the locus defined by $r = 0$. Here

$$\gamma : S^1 \to \mathbf{R}^3 - \{0\}, \ s \mapsto (\gamma_1(s), \ \gamma_2(s), \ \gamma_3(s)), \tag{0.2}$$

is a space curve and (w_1, w_2, w_3) is a triple of three positive integers whose greatest common divisor is one. We assume that γ is an immersion. Observe that we have another space curve

$$\hat{\gamma} : S^1 \to \mathbf{R}^3 - \{0\}, \ s \mapsto ((-1)^{w_1}\gamma_1(s), \ (-1)^{w_2}\gamma_2(s), \ (-1)^{w_3}\gamma_3(s)). \tag{0.3}$$

Then we have two cases to consider.

- If $\gamma(S^1) \neq \hat{\gamma}(S^1)$, then the map g is generically one to one, and we set $\epsilon_g = 1$.
- If $\gamma(S^1) = \hat{\gamma}(S^1)$, we may assume that $\hat{\gamma}(s) = \gamma(s + \pi)$, after suitable re-parametrization of γ, and the map g is generically two to one. We set $\epsilon_g = 2$. We also remark that the map g factors through the natural map $\mathbf{R} \times S^1 \to M$ where M is the Möbius strip obtained by identifying (r, s) with $(-r, s + \pi)$.

The goal of the paper is to give criteria of singularity types of fronts if g is defined by (0.1). In §1, we start with the homogeneous case, that is, Case $(w_1, w_2, w_3) = (1, 1, 1)$, since this case is simple and it is better to treat separately. In §2, we treat Case $(w_1, w_2, w_3) \neq (1, 1, 1)$. To investigate criteria of singularity types of fronts of g, we show a formula of principal curvatures (Theorem 2.3) and describe the behaviors of principal directions near singular locus $\{r = 0\}$ (Theorem 2.9). These information describe how swallowtail singularities of g^t appear and how many of them degenerate to g as $t \to 0$ (see Remark 2.6). In §3, we define focal curves when $(w_1, w_2, w_3) = (1, 2, 2)$. This should be considered as analogy of focal conics of Whitney umbrella which is discussed in [3]. In §4, we discuss some examples. We construct a C^3- singular surface (Example 4.1) so that the principal curvatures are bounded.

1. Fronts of cones

First we consider Case $(w_1, w_2, w_3) = (1, 1, 1)$, that is, a cone defined by

$$g : \mathbf{R} \times S^1 \to \mathbf{R}^3 - \{0\}, \ (r, s) \mapsto r\gamma(s).$$

Without loss of generality we may assume that γ is a spherical curve and s is the arc-length parameter. We then have that

$$\langle \gamma(s), \gamma(s) \rangle \equiv 1, \qquad \langle \gamma'(s), \gamma'(s) \rangle \equiv 1, \qquad \text{and} \qquad \langle \gamma'(s), \gamma(s) \rangle = 0.$$

$\mathbf{a}_1 = \gamma'(s)$, $\mathbf{a}_2 = \gamma(s) \times \gamma'(s)$, $\mathbf{a}_3 = \gamma(s)$ form a frame, and we have

$$\begin{pmatrix} \mathbf{a}_1' \\ \mathbf{a}_2' \\ \mathbf{a}_3' \end{pmatrix} = \begin{pmatrix} 0 & \kappa_g & -1 \\ -\kappa_g & 0 & 0 \\ 1 & 0 & 0 \end{pmatrix} \begin{pmatrix} \mathbf{a}_1 \\ \mathbf{a}_2 \\ \mathbf{a}_3 \end{pmatrix}$$

where $\kappa_g(s)$ is the geodesic curvature. Since

$$g_r = \gamma(s) = \mathbf{a}_3, \qquad g_s = r\gamma'(s) = r\mathbf{a}_1,$$

$\mathbf{a}_2$ is a unit normal to the cone, and the first fundamental form is given by

$$\mathrm{I} = dr^2 + r^2 ds^2.$$

Since

$$g_{rr} = 0, \qquad g_{rs} = \gamma'(s) = \mathbf{a}_1, \qquad g_{ss} = r\gamma''(s) = r(\kappa_g \mathbf{a}_2 - \mathbf{a}_3),$$

the second fundamental form is given by

$$\mathrm{II} = r\kappa_g \, ds^2,$$

and the principal curvatures are 0 and $\frac{\kappa_g}{r}$ with principal directions ∂_r, ∂_s, respectively.

Now we define fronts of the cone by

$$g^t(r, s) = g(r, s) + t\mathbf{a}_2 = r\mathbf{a}_3 + t\mathbf{a}_2.$$

Since

$$\frac{\partial g^t}{\partial r} = \mathbf{a}_3, \qquad \frac{\partial g^t}{\partial s} = (r - t\kappa_g)\mathbf{a}_1,$$

$s \mapsto (t\kappa_g(s), s)$ is the singular curve and $t\kappa_g'(s)\partial_r + \partial_s$ is its tangent (singular direction) and the null direction is ∂_s. Then we have the following criteria, by the criteria of the singularities of front in [4].

- The singularity type of g^t is cuspidal edge at (r_0, s_0), if $r_0 = t\kappa_g(s_0)$, $\kappa_g'(s_0) \neq 0$.
- The singularity type of g^t is swallowtail at (r_0, s_0), if $r_0 = t\kappa_g(s_0)$, $\kappa_g'(s_0) = 0$, $\kappa_g''(s_0) \neq 0$.

2. Weighted cones

We consider the weighted cone defined by (0.1) with $(w_1, w_2, w_3) \neq (1, 1, 1)$. We assume that $w_1 \leq w_2 \leq w_3$.

2.1. Unit normals and fundamental forms

Lemma 2.1. *The unit normal vector* $\mathbf{n} = \frac{g_r \times g_s}{\|g_r \times g_s\|}$ *of the weighted cone* $g(r,s)$ *has the unique extension except the points* $(r,s) = (0,s_0)$ *where*

- $\begin{vmatrix} w_1\gamma_1 & \gamma_1' \\ w_2\gamma_2 & \gamma_2' \end{vmatrix} = 0$ *at* $s = s_0$ *if* $w_2 < w_3$, *or*

- $\left(\begin{vmatrix} w_1\gamma_1 & \gamma_1' \\ w_3\gamma_2 & \gamma_2' \end{vmatrix}, \begin{vmatrix} w_1\gamma_1 & \gamma_1' \\ w_3\gamma_3 & \gamma_3' \end{vmatrix} \right) = (0,0)$ *at* $s = s_0$ *if* $w_2 = w_3$.

Proof. Set $\mathbf{e}_1 = (1,0,0)$, $\mathbf{e}_2 = (0,1,0)$ and $\mathbf{e}_3 = (0,0,1)$. Since

$$g_r(s,r) = (w_1 r^{w_1-1}\gamma_1(s),\ w_2 r^{w_2-1}\gamma_2(s),\ w_3 r^{w_3-1}\gamma_3(s)),$$
$$g_s(s,r) = (r^{w_1}\gamma_1'(s),\ r^{w_2}\gamma_2'(s),\ r^{w_3}\gamma_3'(s)),$$

we obtain

$$g_r \times g_s = \begin{vmatrix} w_1 r^{w_1-1}\gamma_1(s) & r^{w_1}\gamma_1'(s) & \mathbf{e}_1 \\ w_2 r^{w_2-1}\gamma_2(s) & r^{w_2}\gamma_2'(s) & \mathbf{e}_2 \\ w_3 r^{w_3-1}\gamma_3(s) & r^{w_3}\gamma_3'(s) & \mathbf{e}_3 \end{vmatrix}$$

$$= r^{w_1+w_2+w_3-1} \begin{vmatrix} w_1\gamma_1(s) & \gamma_1'(s) & r^{-w_1}\mathbf{e}_1 \\ w_2\gamma_2(s) & \gamma_2'(s) & r^{-w_2}\mathbf{e}_2 \\ w_3\gamma_3(s) & \gamma_3'(s) & r^{-w_3}\mathbf{e}_3 \end{vmatrix}$$

$$= r^{w_1+w_2-1} \begin{vmatrix} w_1\gamma_1(s) & \gamma_1'(s) & r^{w_3-w_1}\mathbf{e}_1 \\ w_2\gamma_2(s) & \gamma_2'(s) & r^{w_3-w_2}\mathbf{e}_2 \\ w_3\gamma_3(s) & \gamma_3'(s) & \mathbf{e}_3 \end{vmatrix}.$$

Then the unit normal vector

$$\mathbf{n} = \frac{g_r \times g_s}{A} = \frac{1}{A_1} \begin{vmatrix} w_1\gamma_1 & \gamma_1' & r^{w_3-w_1}\mathbf{e}_1 \\ w_2\gamma_2 & \gamma_2' & r^{w_3-w_2}\mathbf{e}_2 \\ w_3\gamma_3 & \gamma_3' & \mathbf{e}_3 \end{vmatrix}$$

is extensible near $(r,s) = (0,s_0)$ in $\mathbf{R} \times S^1$ under the assumptions. Here A is defined by

$$A^2 = EG - F^2$$

$$= \det \begin{pmatrix} w_1 r^{w_1-1}\gamma_1 & w_2 r^{w_2-1}\gamma_2 & w_3 r^{w_3-1}\gamma_3 \\ r^{w_1}\gamma_1' & r^{w_2}\gamma_2' & r^{w_3}\gamma_3' \end{pmatrix} \begin{pmatrix} w_1 r^{w_1-1}\gamma_1 & r^{w_1}\gamma_1' \\ w_2 r^{w_2-1}\gamma_2 & r^{w_2}\gamma_2' \\ w_3 r^{w_3-1}\gamma_3 & r^{w_3}\gamma_3' \end{pmatrix}$$

$$= r^{2(w_1+w_2-1)} A_1{}^2,$$

and A_1 is the non-negative function defined by

$$A_1{}^2 = \begin{vmatrix} w_1\gamma_1(s) & \gamma_1'(s) \\ w_2\gamma_2(s) & \gamma_2'(s) \end{vmatrix}^2 + \begin{vmatrix} w_1\gamma_1(s) & \gamma_1'(s) \\ w_3\gamma_3(s) & \gamma_3'(s) \end{vmatrix}^2 r^{2(w_3-w_2)}$$

$$+ \begin{vmatrix} w_2\gamma_2(s) & \gamma_2'(s) \\ w_3\gamma_3(s) & \gamma_3'(s) \end{vmatrix}^2 r^{2(w_3-w_1)}. \qquad \square$$

Lemma 2.2. *The first and second fundamental forms* I, II *for* $g(r,s)$ *are given by the following formulas:*

$$\mathrm{I} = r^{2(w_1-1)}(E_1\,dr^2 + 2F_1\,r\,dr\,ds + G_1\,r^2ds^2),$$

$$\mathrm{II} = \frac{r^{w_3-2}}{A_1}(L_0\,dr^2 + 2M_0\,r\,dr\,ds + N_0\,r^2ds^2),$$

where

$$E_1 = (w_1\gamma_1)^2 + r^{2(w_2-w_1)}(w_2\gamma_2)^2 + r^{2(w_3-w_1)}(w_3\gamma_3)^2,$$

$$F_1 = w_1\gamma_1\gamma_1' + r^{2(w_2-w_1)}w_2\gamma_2\gamma_2' + r^{2(w_3-w_1)}w_3\gamma_3\gamma_3',$$

$$G_1 = (\gamma_1')^2 + r^{2(w_2-w_1)}(\gamma_2')^2 + r^{2(w_3-w_1)}(\gamma_3')^2,$$

$$L_0 = \begin{vmatrix} w_1\gamma_1 & \gamma_1' & w_1{}^2\gamma_1 \\ w_2\gamma_2 & \gamma_2' & w_2{}^2\gamma_2 \\ w_3\gamma_3 & \gamma_3' & w_3{}^2\gamma_3 \end{vmatrix}, \quad M_0 = \begin{vmatrix} w_1\gamma_1 & \gamma_1' & w_1\gamma_1' \\ w_2\gamma_2 & \gamma_2' & w_2\gamma_2' \\ w_3\gamma_3 & \gamma_3' & w_3\gamma_3' \end{vmatrix}, \quad N_0 = \begin{vmatrix} w_1\gamma_1 & \gamma_1' & \gamma_1'' \\ w_2\gamma_2 & \gamma_2' & \gamma_2'' \\ w_3\gamma_3 & \gamma_3' & \gamma_3'' \end{vmatrix}.$$

Proof. The first fundamental form I is given by

$$\mathrm{I} = E\,dr^2 + 2F\,dr\,ds + G\,ds^2$$

where

$$E = w_1^2 r^{2w_1-2}(\gamma_1)^2 + w_2^2 r^{2w_2-2}(\gamma_2)^2 + w_3^2 r^{2w_3-2}(\gamma_3)^2 = r^{2w_1-2}E_1,$$

$$F = w_1 r^{2w_1-1}\gamma_1\gamma_1' + w_2 r^{2w_2-1}\gamma_2\gamma_2' + w_3 r^{2w_3-1}\gamma_3\gamma_3' = r^{2w_1-1}F_1,$$

$$G = r^{2w_1}\gamma_1^2 r^{2w_2}\gamma_2^2 + r^{2w_3}\gamma_3^2 = r^{2w_1}G_1.$$

Since

$$g_{rr}(r,s) = (w_1(w_1-1)r^{w_1-2}\gamma_1(s),\ w_2(w_2-1)r^{w_2-2}\gamma_2(s),\ w_3(w_3-1)r^{w_3-2}\gamma_3(s)),$$

$$g_{rs}(s,r) = (w_1 r^{w_1-1}\gamma_1'(s),\ w_2 r^{w_2-1}\gamma_2'(s),\ w_3 r^{w_3-1}\gamma_3'(s)),$$

$$g_{ss}(s,r) = (r^{w_1}\gamma_1''(s),\ r^{w_2}\gamma_2''(s),\ r^{w_3}\gamma_3''(s)),$$

we have

$$L = \langle \mathbf{n}, g_{rr}\rangle = \frac{r^{w_3-2}}{A_1}L_0,$$

$$M = \langle \mathbf{n}, g_{rs} \rangle = \frac{r^{w_3 - 1}}{A_1} M_0,$$

$$N = \langle \mathbf{n}, g_{ss} \rangle = \frac{r^{w_3}}{A_1} N_0,$$

and we obtain the formula for the second fundamental form. $\qquad\square$

2.2. Curvatures of weighted cones

Theorem 2.3. *The Gaussian curvature K and the mean curvature H of g are given by*

$$K = \frac{L_0 N_0 - M_0^2}{A_1^4} r^{2(w_3 - w_1 - w_2)}, \qquad H = \frac{H_1}{2A_1^3} r^{w_3 - 2w_1},$$

where $H_1 = E_1 N_0 - 2F_1 M_0 + G_1 L_0$. Moreover we have the following expansion for principal curvatures. If $w_1 < w_2$, then

$$\kappa_1 = \frac{L_0 N_0 - M_0^2}{A_1 H_1} r^{w_3 - 2w_1} \left(1 + \frac{A_1^2 (L_0 N_0 - M_0^2)}{H_1^2} r^{2w_2 - 2w_1} + \cdots \right),$$

$$\kappa_2 = \frac{H_1}{A_1^3} r^{w_3 - 2w_2} \left(1 - \frac{A_1^2 (L_0 N_0 - M_0^2)}{H_1^2} r^{2w_2 - 2w_1} + \cdots \right),$$

where "$\cdots$" denotes terms of $4(w_2 - w_1)$ and higher order in r. If $w_1 = w_2$, then

$$\kappa_1 = \frac{H_1 - \sqrt{H_1^2 - 4A_1^2 (L_0 N_0 - M_0^2)}}{2A_1^3} r^{w_3 - 2w_1},$$

$$\kappa_2 = \frac{H_1 + \sqrt{H_1^2 - 4A_1^2 (L_0 N_0 - M_0^2)}}{2A_1^3} r^{w_3 - 2w_1}.$$

Proof. The Gaussian curvature K is expressed as

$$K = \frac{LN - M^2}{EG - F^2} = \frac{L_0 N_0 - M_0^2}{A^2 A_1^2} r^{2(w_3 - 1)} = \frac{L_0 N_0 - M_0^2}{A_1^4} r^{2(w_3 - w_1 - w_2)}.$$

Since $E = E_1 r^{2w_1 - 2}$, $F = F_1 r^{2w_1 - 1}$, $G = G_1 r^{2w_1}$, $L = \frac{L_0}{A_1} r^{w_3 - 2}$, $M = \frac{M_0}{A_1} r^{w_3 - 1}$, $N = \frac{N_0}{A_1} r^{w_3}$, $A^2 = r^{2w_1 + 2w_2 - 2} A_1^2$, we obtain that

$$H = \frac{EN - 2FM + GL}{2(EG - F^2)} = \frac{H_1}{2A_1^3} r^{w_3 - 2w_2}.$$

Therefore the principal curvatures are expressed as follows:

$$\kappa = H \pm \sqrt{H^2 - K}$$

$$= \frac{H_1}{2A_1^3} r^{w_3-2w_2} \pm \sqrt{\frac{H_1^2}{4A_1^6} r^{2w_3-4w_2} - \frac{L_0 N_0 - M_0^2}{A_1^4} r^{2w_3-2w_2-2w_1}}$$

$$= \frac{r^{w_3-2w_2}}{2A_1^3} \left(H_1 \pm \sqrt{H_1^2 - 4A_1^2(L_0 N_0 - M_0^2) r^{2w_2-2w_1}} \right)$$

$$= \frac{r^{w_3-2w_2} H_1}{2A_1^3} \left(1 \pm \sqrt{1 - \frac{4A_1^2(L_0 N_0 - M_0^2)}{H_1^2} r^{2w_2-2w_1}} \right)$$

$$= \frac{r^{w_3-2w_2} H_1}{2A_1^3} R$$

where $R = 1 \pm (1 - \frac{2A_1^2(L_0 N_0 - M_0^2)}{H_1^2} r^{2w_2-2w_1} - \frac{2A_1^4(L_0 N_0 - M_0^2)^2}{H_1^4} r^{4w_2-4w_1} + \cdots)$. We thus obtain the result for κ_i. $\qquad\square$

2.3. *Ridge points, subparabolic points and fronts of weighted cones*

For a regular surface with principal curvatures κ_i and principal directions v_i, a ridge point relative to v_i is defined by a point satisfying $v_i \kappa_i = 0$. An m-th order ridge point relative to v_i is defined by a point with $v_i \kappa_i = v_i^2 \kappa_i = \cdots = v_i^m \kappa_i = 0$, $v_i^{m+1} \kappa_i \neq 0$. A subparabolic point relative to v_i is defined by a point with $v_i \kappa_j = 0$ for $i \neq j$. We are able to generalize these definitions for weighted cones if κ_i is extensible over the singularity.

Definition 2.4. Set $i = 1, 2$ and assume that $w_3 = 2w_i$. Let $\tilde{v}_i$ denote a vector field so that $r^d v_i = \tilde{v}_i$ and $\tilde{v}_i|_{r=0} \neq 0$. We say that a point $(r, s) = (0, s_0)$ is a **ridge point** relative to v_i if $\tilde{v}_i \kappa_i = 0$ at $(r, s) = (0, s_0)$. We also say a point $(r, s) = (0, s_0)$ is an **m-th order ridge point** relative to v_i if $\tilde{v}_i \kappa_i = \tilde{v}_i \kappa_i = \tilde{v}_i^2 \kappa_i = \cdots = \tilde{v}_i^m \kappa_i = 0$ and $\tilde{v}_i^{m+1} \kappa_i \neq 0$ at $(r, s) = (0, s_0)$. We say a point $(r, s) = (0, s_0)$ is a **subparabolic point** relative to v_i if $\tilde{v}_i \kappa_j = 0$ at $(r, s) = (0, s_0)$ where $j = 1, 2$ with $i \neq j$.

Now we define the fronts of weighted cone by

$$g^t : \mathbf{R} \times S \to \mathbf{R}^3, \quad g^t(r, s) = g(r, s) + t\mathbf{n}(r, s),$$

where

$$S = \begin{cases} \left\{ s \in S^1 : \begin{vmatrix} w_1 \gamma_1 & \gamma_1' \\ w_2 \gamma_2 & \gamma_2' \end{vmatrix} \neq 0 \right\} & (w_2 < w_3) \\ \left\{ s \in S^1 : \left(\begin{vmatrix} w_1 \gamma_1 & \gamma_1' \\ w_2 \gamma_2 & \gamma_2' \end{vmatrix}, \begin{vmatrix} w_1 \gamma_1 & \gamma_1' \\ w_2 \gamma_3 & \gamma_3' \end{vmatrix} \right) \neq (0, 0) \right\} & (w_2 = w_3) \end{cases}.$$

Set $\Sigma_k = \Sigma_k^{(1)} \cup \Sigma_k^{(2)}$ where $\Sigma_k^{(i)} = \{(r,s) \in \mathbf{R} \times S^1 : \kappa_i(r,s) = k\}$. It is well-known that Σ_k is the singular locus of g^t for $t = 1/k$.

Theorem 2.5.

(1) If $(r_0, s_0) \in \Sigma_{1/t}^{(i)}$ is not ridge relative to v_i, then g^t has cuspidal edge singularity at (r_0, s_0).

(2) If $(r,s) = (r_0, s_0) \in \Sigma_{1/t}^{(i)}$ is a first order ridge point relative to v_i and not subparabolic relative to v_j $(j \neq i)$, then g^t has swallowtail singularity at (r_0, s_0).

The proof is similar to that appeared in [3, Theorem 4.1], and we omit the details. Define $R_{i,0}(s)$ $(i = 1, 2)$ by

$$\tilde{v}_i \kappa_i = r^{w_3 - 2w_i}(R_{i,0}(s) + O(r^2)).$$

Remark 2.6. If the equation $R_{i,0}(s) = 0$ has r_i simple solutions on the set S, then there is r_i swallowtail singularities of g^t for $t = 1/k$ near $r = 0$. So we observe that

- If $w_3 < 2w_1$, $\epsilon_g(r_1 + r_2)$ number of swallowtail singularities of the map g^t tend to the origin as $t \to 0$.
- If $2w_1 < w_3 < 2w_2$, $\epsilon_g r_2$ number of swallowtail singularities of the map g^t tend to the origin as $t \to 0$.

To show a closed formula for $R_{i,0}$, we introduce the following notations:

$$a_0 = \begin{vmatrix} w_1\gamma_1 & \gamma_1' \\ w_2\gamma_2 & \gamma_2' \end{vmatrix}^2, \qquad a_1 = \begin{vmatrix} w_1\gamma_1 & \gamma_1' \\ w_3\gamma_3 & \gamma_3' \end{vmatrix}^2, \qquad a_2 = \begin{vmatrix} w_2\gamma_2 & \gamma_2' \\ w_3\gamma_3 & \gamma_3' \end{vmatrix}^2,$$

$$h_0 = \begin{vmatrix} w_1\gamma_1 & \gamma_1' & w_1{}^2\gamma_1(\gamma_1\gamma_1'' - \gamma_1'^2) \\ w_2\gamma_2 & \gamma_2' & w_2{}^2\gamma_2\gamma_1'^2 - 2w_1w_2\gamma_1\gamma_1'\gamma_2' + w_1{}^2\gamma_1{}^2\gamma_2'' \\ w_3\gamma_3 & \gamma_3' & w_3{}^2\gamma_3\gamma_1'^2 - 2w_1w_3\gamma_1\gamma_1'\gamma_3' + w_1{}^2\gamma_1{}^2\gamma_3'' \end{vmatrix},$$

$$h_1 = \begin{vmatrix} w_1\gamma_1 & \gamma_1' & w_1{}^2\gamma_1\gamma_2'^2 - 2w_1w_2\gamma_2\gamma_1'\gamma_2' + w_2{}^2\gamma_2{}^2\gamma_1'' \\ w_2\gamma_2 & \gamma_2' & w_2{}^2\gamma_2(\gamma_2\gamma_2'' - \gamma_2'^2) \\ w_3\gamma_3 & \gamma_3' & w_3{}^2\gamma_3\gamma_2'^2 - 2w_2w_3\gamma_2\gamma_2'\gamma_3' + w_2{}^2\gamma_2{}^2\gamma_3'' \end{vmatrix},$$

$$h_2 = \begin{vmatrix} w_1\gamma_1 & \gamma_1' & w_1{}^2\gamma_1\gamma_3'^2 - 2w_1w_3\gamma_3\gamma_1'\gamma_3' + w_3{}^2\gamma_3{}^2\gamma_1'' \\ w_2\gamma_2 & \gamma_2' & w_2{}^2\gamma_2\gamma_3'^2 - 2w_2w_3\gamma_3\gamma_2'\gamma_3' + w_3{}^2\gamma_3{}^2\gamma_2'' \\ w_3\gamma_3 & \gamma_3' & w_3{}^2\gamma_3(\gamma_3\gamma_3'' - \gamma_3'^2) \end{vmatrix}.$$

Proposition 2.7. We have the following closed formulas for $R_{i,0}$.

(1) If $w_1 < w_2 < w_3$, then we have

$$R_{1,0} = (w_3 - 2w_1)(N_0 - \frac{(\gamma_1')^2(L_0 N_0 - M_0{}^2)}{h_0})\frac{L_0 N_0 - M_0{}^2}{a_0{}^{1/2} h_0}$$
$$- (M_0 - \frac{\gamma_1' \gamma_1 (L_0 N_0 - M_0{}^2)}{h_0})(\frac{L_0 N_0 - M_0{}^2}{a_0{}^{1/2} h_0})',$$

$$R_{2,0} = (w_3 - 2w_2)\gamma_1' \frac{h_0}{a_0{}^{3/2}} - w_1 \gamma_1 (\frac{h_0}{a_0{}^{3/2}})'.$$

(2) If $w_1 = w_2 < w_3$, we have

$$R_{1,0} = (w_3 - 2w_1)(N_0 - \frac{((\gamma_1')^2 + (\gamma_2')^2)(L_0 N_0 - M_0{}^2)}{h_0 + h_1})\frac{L_0 N_0 - M_0{}^2}{a_0{}^{1/2}(h_0 + h_1)}$$
$$- (M_0 - \frac{w_1(\gamma_1'\gamma_1 + \gamma_2'\gamma_2)(L_0 N_0 - M_0{}^2)}{h_0 + h_1})(\frac{L_0 N_0 - M_0{}^2}{a_0{}^{1/2}(h_0 + h_1)})',$$

$$R_{2,0} = (w_3 - 2w_1)\left(N_0 - \frac{((\gamma_1')^2 + (\gamma_2')^2)(L_0 N_0 - M_0{}^2)}{h_0 + h_1}\right)\frac{h_0 + h_1}{a_0{}^{3/2}}$$
$$- (M_0 - \frac{w_1(\gamma_1'\gamma_1 + \gamma_2'\gamma_2)(L_0 N_0 - M_0{}^2)}{h_0 + h_1})(\frac{h_0 + h_1}{a_0{}^{3/2}})'.$$

(3) If $w_1 < w_2 = w_3$, then we have

$$R_{1,0} = (w_3 - 2w_1)\left(N_0 + \frac{w_3(w_3 - w_1)(\gamma_1')^2}{w_1 \gamma_1}\begin{vmatrix} \gamma_2 & \gamma_2' \\ \gamma_3 & \gamma_3' \end{vmatrix}\right)\frac{h_0}{(a_0 + a_1)^{3/2}},$$

$$R_{2,0} = -w_3(N_0 - \frac{(\gamma_1')^2 h_0}{w_1{}^2(\gamma_1 \gamma_2' - \gamma_1' \gamma_2)^2})\frac{h_0}{(a_0 + a_1)^{3/2}}$$
$$- (M_0 - \frac{\gamma_1 \gamma_1' h_0}{w_1(\gamma_1 \gamma_2' - \gamma_1' \gamma_2)^2})(\frac{h_0}{(a_0 + a_1)^{3/2}})'.$$

The proof of the proposition will be given in the next subsection.

Corollary 2.8. *Assume that $(w_1, w_2, w_3) = (1, 2, 2)$. Then the locus defined by $r = 0$ is in the ridge locus relative to v_1 and the singularity types of g^t on this locus are not cuspidal edge.*

Proof. By the previous proposition, we have $R_{1,0} = 0$. This implies the result. □

2.4. *Principal directions of weighted cones*

Theorem 2.9. *Principal directions v_i are given by*

$$(N - \kappa_i G)\partial_r - (M - \kappa_i F)\partial_s.$$

The order of each components of v_i in r is given by the following table:

	$N - \kappa_1 G$	$M - \kappa_1 F$	$N - \kappa_2 G$	$M - \kappa_2 F$
$w_2 < w_3$	w_3	$w_3 - 1$	$w_3 + 2w_1 - 2w_2$	$w_3 + 2w_1 - 2w_2 - 1$
$w_2 = w_3$	w_3	$3w_3 - 2w_1 - 1$	$2w_1 - w_3$	$2w_1 - w_3 - 1$

This theorem asserts that the limit of principal direction which is represented by v_1 is ∂_r as $r \to 0$ for generic s, when $w_1 < w_2 = w_3$. In the other cases, the limits of principal directions are ∂_s for generic s.

Proof. Since

$$N - \kappa_1 G = r^{w_3} \frac{N_0}{A_1}$$

$$- \frac{L_0 N_0 - M_0{}^2}{A_1 H_1} r^{w_3 - 2w_1} \left(1 + \frac{A_1{}^2(L_0 N_0 - M_0{}^2)}{H_1{}^2} r^{2w_2 - 2w_1} + \cdots \right) G_1 r^{2w_1}$$

$$= \frac{G_1}{A_1} r^{w_3} \left(\frac{N_0}{G_1} - \frac{L_0 N_0 - M_0{}^2}{H_1} (1 + \frac{A_1{}^2(L_0 N_0 - M_0{}^2)}{H_1{}^2} r^{2w_2 - 2w_1} + \cdots) \right),$$

$$M - \kappa_1 F = r^{w_3 - 1} \frac{M_0}{A_1}$$

$$- r^{w_3 - 2w_1} \frac{L_0 N_0 - M_0{}^2}{A_1 H_1} r^{w_3 - 2w_1} \left(1 + \frac{A_1{}^2(L_0 N_0 - M_0{}^2)}{H_1{}^2} r^{2w_2 - 2w_1} + \cdots \right) F_1 r^{2w_1 - 1}$$

$$= \frac{F_1}{A_1} r^{w_3 - 1} \left(\frac{M_0}{F_1} - \frac{L_0 N_0 - M_0{}^2}{H_1} (1 + \frac{A_1{}^2(L_0 N_0 - M_0{}^2)}{H_1{}^2} r^{2w_2 - 2w_1} + \cdots) \right),$$

$$N - \kappa_2 G = r^{w_3} \frac{N_0}{A_1}$$

$$- \frac{H_1}{A_1{}^3} r^{w_3 - 2w_2} \left(1 - \frac{A_1{}^2(L_0 N_0 - M_0{}^2)}{H_1{}^2} r^{2w_2 - 2w_1} + \cdots \right) G_1 r^{2w_1}$$

$$= r^{w_3} \frac{N_0}{A_1} - \frac{H_1 G_1}{A_1{}^3} r^{w_3 + 2w_1 - 2w_2} \left(1 - \frac{A_1{}^2(L_0 N_0 - M_0{}^2)}{H_1{}^2} r^{2w_2 - 2w_1} + \cdots \right),$$

$$M - \kappa_2 F = r^{w_3 - 1} \frac{M_0}{A_1}$$

$$- \frac{H_1}{A_1{}^3} r^{w_3 - 2w_2} \left(1 - \frac{A_1{}^2(L_0 N_0 - M_0{}^2)}{H_1{}^2} r^{2w_2 - 2w_1} + \cdots \right) F_1 r^{2w_1 - 1}$$

$$= r^{w_3 - 1} \frac{M_0}{A_1} - \frac{H_1 F_1}{A_1{}^3} r^{w_3 + 2w_1 - 2w_2 - 1} \left(1 - \frac{A_1{}^2(L_0 N_0 - M_0{}^2)}{H_1{}^2} r^{2w_2 - 2w_1} + \cdots \right),$$

we obtain that

$$v_1 = (N - \kappa_1 G)\partial_r - (M - \kappa_1 F)\partial_s$$

$$= \frac{r^{w_3 - 1}}{A_1} \left[\left(N_0 - \frac{G_1(L_0 N_0 - M_0{}^2)}{H_1} + O(r^2)\right) r \partial_r \right.$$

$$\left. - \left(M_0 - \frac{F_1(L_0 N_0 - M_0{}^2)}{H_1} + O(r^2)\right) \partial_s \right],$$

$$v_2 = (N - \kappa_2 G)\partial_r - (M - \kappa_2 F)\partial_s$$

$$= \begin{cases} \frac{H_1 \gamma_1'}{A_1{}^3} r^{w_3 + 2w_1 - 2w_2 - 1} \left[(\gamma_1' + O(r^2)) r \partial_r - (w_1 \gamma_1 + O(r^2))\partial_s \right] & (w_1 < w_2) \\ \frac{r^{w_3 - 1}}{A_1} \left[\left(N_0 - \frac{G_1 H_1}{A_1{}^2} + O(r^2)\right) r \partial_r - \left(M_0 - \frac{F_1 H_1}{A_1{}^2} + O(r^2)\right)\partial_s \right] & (w_1 = w_2) \end{cases}.$$

Since

$$\begin{vmatrix} L_0 & M_0 \\ w_1\gamma_1 & \gamma_1' \end{vmatrix} = (w_2 - w_3)\begin{vmatrix} w_1\gamma_1 & \gamma_1' \\ w_2\gamma_2 & \gamma_2' \end{vmatrix}\begin{vmatrix} w_1\gamma_1 & \gamma_1' \\ w_3\gamma_3 & \gamma_3' \end{vmatrix},$$

we observe ∂_r is a principal direction when $w_2 = w_3$ and $r = 0$. This suggests a cancellation of the coefficient of r^{w_3-1} in $M - \kappa_1 F$. In fact, we have

$$\begin{vmatrix} M_0 & L_0 N_0 - M_0{}^2 \\ \gamma_1' & H_1|_{r=0} \end{vmatrix} = (w_3 - w_2)\begin{vmatrix} w_1\gamma_1 & \gamma_1' \\ w_2\gamma_2 & \gamma_2' \end{vmatrix}\begin{vmatrix} w_1\gamma_1 & \gamma_1' \\ w_3\gamma_3 & \gamma_3' \end{vmatrix}\begin{vmatrix} w_1\gamma_1 & M_0 \\ \gamma_1' & N_0 \end{vmatrix},$$

and we see a cancellation of the coefficient of r^{w_3-1} in $M - \kappa_1 F$, when $w_2 = w_3$.

When $w_1 < w_2 = w_3$,

$$\begin{vmatrix} N_0 & L_0 N_0 - M_0{}^2 \\ G_1|_{r=0} & H_1|_{r=0} \end{vmatrix}$$

$$= \begin{vmatrix} M_0 & w_1\gamma_1 \\ N_0 & \gamma_1' \end{vmatrix}\left(\begin{vmatrix} M_0 & w_1\gamma_1 \\ N_0 & \gamma_1' \end{vmatrix} + (w_1 - w_3)w_3(\gamma_1'' - \gamma_1'{}^2)\begin{vmatrix} \gamma_2' & \gamma_2 \\ \gamma_3' & \gamma_3 \end{vmatrix} \right).$$

This implies that the initial coefficient of $N - \kappa_1 F$ is not identically zero. In this way, we obtain the table desired. $\qquad\square$

Remark 2.10. Set $\kappa = r^e(k_0 + k_2 r^2 + \cdots)$.

- When $v = (\alpha_1 r + \alpha_3 r^3 + \cdots)\partial_r + (\beta_0 + \beta_2 r^2 + \cdots)\partial_s$, we have

$$\begin{aligned} v\kappa &= (\alpha_1 r + \alpha_3 r^3 + \cdots)(ek_0 r^{e-1} + (e+2)k_2 r^{e+1} + \cdots) \\ &\quad + (\beta_0 + \beta_2 r^2 + \cdots)r^e(k_0' + k_2' r^2 + \cdots) \\ &= (e\alpha_1 k_0 + \beta_0 k_0')r^e + O(r^{e+2}), \\ v^2\kappa &= \left(e\alpha_1(e\alpha_1 k_0 + \beta_0 k_0') + (e\alpha_1 k_0 + \beta_0 k_0')' \right)r^e + O(r^{e+2}). \end{aligned}$$

If $e = 0$, then we obtain

$$\begin{aligned} v\kappa &= \beta_0 k_0' + O(r^2), \\ v^2\kappa &= (2\beta_0 k_0' + \beta_0' k_0) + O(r^2). \end{aligned}$$

- When $v = (\alpha_0 + \alpha_2 r^2 + \cdots)\partial_r + (\beta_1 r + \beta_3 r^3 + \cdots)\partial_s$, we have

$$\begin{aligned} v\kappa &= (\alpha_0 + \alpha_2 r^2 + \cdots)(ek_0 r^{e-1} + (e+2)k_2 r^{e+1} + \cdots \\ &\quad + (\beta_1 r + \beta_3 r^3 + \cdots)r^e(k_0' + k_2' r^2 + \cdots) \\ &= e\alpha_0 k_0 r^{e-1} + (\alpha_0(e+2)k_2 + \alpha_2 ek_0 + \beta_1 k_0')r^{e+1} + O(r^{e+2}), \end{aligned}$$

$$v^2\kappa = \alpha_0(e(e-1)\alpha_0 k_0 r^{e-2}$$
$$+ (e+1)(\alpha_0(e+2)k_2 + \alpha_2 e k_0 + \beta_1 k_0')r^e + \cdots)$$
$$+ \beta_1(e(\alpha_0 k_0)'r^e + \cdots)$$
$$= e(e-1)\alpha_0{}^2 k_0 r^{e-2} + ((e+1)\alpha_0(\alpha_0(e+2)k_2 + \alpha_2 e k_0 + \beta_1 k_0')$$
$$+ \beta_1 e(\alpha_0 k_0)')r^e + \cdots .$$

If $e = 0$, then we obtain

$$v\kappa = (2\alpha_0 k_2 + \beta_1 k_0')r + O(r^2),$$
$$v^2\kappa = \alpha_0(2\alpha_0 k_2 + \beta_1 k_0') + O(r^2).$$

We describe below the expression of $\tilde{v}_i \kappa_j$ for $i, j = 1, 2$. These complete the proof of Proposition 2.7, since they contain the expansions of $\tilde{v}_i \kappa_i$ $(i = 1, 2)$.

- When $w_1 < w_2 < w_3$, we have

$$\kappa_1 = \frac{L_0 N_0 - M_0{}^2}{a_0{}^{1/2} h_0} r^{w_3 - 2w_1}(1 + O(r^2)),$$

$$\kappa_2 = \frac{h_0}{a_0{}^{3/2}} r^{w_3 - 2w_2}(1 + O(r^2)),$$

$$\tilde{v}_1 = (N_0 - \tfrac{(\gamma_1')^2(L_0 N_0 - M_0{}^2)}{h_0} + O(r^2))r\partial_r$$
$$- (M_0 - \tfrac{\gamma_1'\gamma_1(L_0 N_0 - M_0{}^2)}{h_0} + O(r^2))\partial_s,$$

$$\tilde{v}_2 = (\gamma_1' + O(r^2))r\partial_r - (w_1\gamma_1 + O(r^2))\partial_s.$$

Then we conclude that

$$\tilde{v}_1\kappa_1 = r^{w_3 - 2w_1}\Big[(w_3 - 2w_1)(N_0 - \tfrac{(\gamma_1')^2(L_0 N_0 - M_0{}^2)}{h_0})\frac{L_0 N_0 - M_0{}^2}{a_0{}^{1/2} h_0}$$
$$- (M_0 - \tfrac{\gamma_1'\gamma_1(L_0 N_0 - M_0{}^2)}{h_0})(\frac{L_0 N_0 - M_0{}^2}{a_0{}^{1/2} h_0})' + O(r^2)\Big],$$

$$\tilde{v}_2\kappa_1 = r^{w_3 - 2w_2}\Big[(w_3 - 2w_2)\gamma_1'\frac{L_0 N_0 - M_0{}^2}{a_0{}^{1/2} h_0} - w_1\gamma_1(\frac{L_0 N_0 - M_0{}^2}{a_0{}^{1/2} h_0})' + O(r^2)\Big],$$

$$\tilde{v}_1\kappa_2 = r^{w_3 - 2w_1}\Big[(w_3 - 2w_1)(N_0 - \tfrac{(\gamma_1')^2(L_0 N_0 - M_0{}^2)}{h_0})\frac{h_0}{a_0{}^{3/2}}$$
$$- (M_0 - \tfrac{\gamma_1'\gamma_1(L_0 N_0 - M_0{}^2)}{h_0})(\frac{h_0}{a_0{}^{3/2}})' + O(r^2)\Big],$$

$$\tilde{v}_2\kappa_2 = r^{w_3 - 2w_2}\Big[(w_3 - 2w_2)\gamma_1'\frac{h_0}{a_0{}^{3/2}} - w_1\gamma_1(\frac{h_0}{a_0{}^{3/2}})' + O(r^2)\Big].$$

- When $w_1 = w_2 < w_3$, we have

$$E_1 = w_1{}^2(\gamma_1{}^2 + \gamma_2{}^2) + r^{2w_3 - 2w_1}w_3{}^2\gamma_3{}^2,$$
$$F_1 = w_1(\gamma_1\gamma_1' + \gamma_2\gamma_2') + r^{2w_3 - 2w_1}w_3\gamma_3\gamma_3',$$

$$G_1 = (\gamma_1')^2 + (\gamma_2')^2 + r^{2w_3-2w_1}(\gamma_3')^2,$$

$$A_1{}^2 = w_1{}^2(\gamma_1\gamma_2' - \gamma_1'\gamma_2)^2 + r^{2w_3-2w_1}((w_1\gamma_1\gamma_3' - w_3\gamma_1'\gamma_3)^2$$
$$+ (w_1\gamma_2\gamma_3' - w_3\gamma_2'\gamma_3)^2),$$

$$L_0 = w_1 w_3 (w_1 - w_3)\gamma_3 \begin{vmatrix} \gamma_1' & \gamma_1 \\ \gamma_2' & \gamma_2 \end{vmatrix},$$

$$M_0 = w_1(w_3 - w_1)\gamma_3' \begin{vmatrix} \gamma_1' & \gamma_1 \\ \gamma_2' & \gamma_2 \end{vmatrix},$$

$$N_0 = \begin{vmatrix} w_1\gamma_1 & \gamma_1' & \gamma_1'' \\ w_1\gamma_2 & \gamma_2' & \gamma_2'' \\ w_3\gamma_3 & \gamma_3' & \gamma_3'' \end{vmatrix}.$$

So we obtain that

$$\kappa_1 = \frac{L_0 N_0 - M_0{}^2}{a_0{}^{1/2}(h_0 + h_1)} r^{w_3-2w_1}(1 + O(r^2)),$$

$$\kappa_2 = \frac{h_0 + h_1}{a_0{}^{3/2}} r^{w_3-2w_2}(1 + O(r^2)),$$

$$\tilde{v}_1 = \Big(N_0 - \tfrac{((\gamma_1')^2+(\gamma_2')^2)(L_0 N_0 - M_0{}^2)}{h_0+h_1} + O(r^2)\Big)r\partial_r$$
$$- (M_0 - \tfrac{w_1(\gamma_1'\gamma_1+\gamma_2'\gamma_2)(L_0 N_0 - M_0{}^2)}{h_0+h_1} + O(r^2))\partial_s,$$

$$\tilde{v}_2 = \big(N_0 - \tfrac{((\gamma_1')^2+(\gamma_2')^2)h_0}{w_1{}^2(\gamma_1\gamma_2'-\gamma_1'\gamma_2)^2} + O(r^2)\big)r\partial_r$$
$$- (M_0 - \tfrac{(\gamma_1\gamma_1'+\gamma_2\gamma_2')h_0}{w_1(\gamma_1\gamma_2'-\gamma_1'\gamma_2)^2} + O(r^2))\partial_s.$$

Then we conclude that

$$\tilde{v}_1\kappa_1 = r^{w_3-2w_2}\Big[(w_3 - 2w_1)(N_0 - \tfrac{((\gamma_1')^2+(\gamma_2')^2)(L_0 N_0-M_0{}^2)}{h_0+h_1})\tfrac{L_0 N_0 - M_0{}^2}{a_0{}^{1/2}(h_0+h_1)}$$
$$- (M_0 - \tfrac{w_1(\gamma_1'\gamma_1+\gamma_2'\gamma_2)(L_0 N_0-M_0{}^2)}{h_0+h_1})\big(\tfrac{L_0 N_0 - M_0{}^2}{a_0{}^{1/2}(h_0 + h_1)}\big)' + O(r^2)\Big],$$

$$\tilde{v}_2\kappa_1 = r^{w_3-2w_2}\Big[(w_3 - 2w_1)(N_0 - \tfrac{((\gamma_1')^2+(\gamma_2')^2)h_0}{w_1{}^2(\gamma_1\gamma_2'-\gamma_1'\gamma_2)^2})\tfrac{h_0+h_1}{a_0{}^{3/2}}$$
$$- (M_0 - \tfrac{(\gamma_1\gamma_1'+\gamma_2\gamma_2')h_0}{w_1(\gamma_1\gamma_2'-\gamma_1'\gamma_2)^2})\big(\tfrac{h_0+h_1}{a_0{}^{3/2}}\big)'\partial_s + O(r^2)\Big],$$

$$\tilde{v}_1\kappa_2 = r^{w_3-2w_2}\Big[(w_3 - 2w_1)(N_0 - \tfrac{((\gamma_1')^2+(\gamma_2')^2)(L_0 N_0-M_0{}^2)}{h_0+h_1})\tfrac{L_0 N_0 - M_0{}^2}{a_0{}^{1/2}h_0}$$
$$- (M_0 - \tfrac{w_1(\gamma_1'\gamma_1+\gamma_2'\gamma_2)(L_0 N_0-M_0{}^2)}{h_0+h_1})\big(\tfrac{L_0 N_0 - M_0{}^2}{a_0{}^{1/2}h_0}\big)' + O(r^2)\Big],$$

$$\tilde{v}_2\kappa_2 = r^{w_3-2w_2}\Big[(w_3 - 2w_1)(N_0 - \tfrac{((\gamma_1')^2+(\gamma_2')^2)(L_0 N_0-M_0{}^2)}{h_0+h_1})\tfrac{h_0+h_1}{a_0{}^{3/2}}$$
$$- (M_0 - \tfrac{w_1(\gamma_1'\gamma_1+\gamma_2'\gamma_2)(L_0 N_0-M_0{}^2)}{h_0+h_1})\big(\tfrac{h_0+h_1}{a_0{}^{3/2}}\big)' + O(r^2)\Big].$$

- When $w_1 < w_2 = w_3$, we have

$$E_1 = (w_1\gamma_1)^2 + r^{2w_3-2w_1}w_3{}^2(\gamma_2{}^2 + \gamma_3{}^2),$$

$$F_1 = w_1\gamma_1\gamma_1' + r^{2w_3-2w_1}w_3(\gamma_2\gamma_2' + \gamma_3\gamma_3'),$$

$$G_1 = (\gamma_1')^2 + r^{2w_3-2w_1}((\gamma_2')^2 + (\gamma_3')^2),$$

$$A_1{}^2 = (w_1\gamma_1\gamma_2' - w_3\gamma_1'\gamma_2)^2 + (w_1\gamma_1\gamma_3' - w_3\gamma_1'\gamma_3)^2$$
$$+ r^{2w_3-2w_1}w_3{}^2(\gamma_2\gamma_3' - \gamma_2'\gamma_3)^2,$$

$$L_0 = w_1w_3(w_3 - w_1)\gamma_1 \begin{vmatrix} \gamma_2' & \gamma_2 \\ \gamma_3' & \gamma_3 \end{vmatrix},$$

$$M_0 = w_3(w_3 - w_1)\gamma_1' \begin{vmatrix} \gamma_2' & \gamma_2 \\ \gamma_3' & \gamma_3 \end{vmatrix},$$

$$N_0 = \begin{vmatrix} w_1\gamma_1 & \gamma_1' & \gamma_1'' \\ w_3\gamma_2 & \gamma_2' & \gamma_2'' \\ w_3\gamma_3 & \gamma_3' & \gamma_3'' \end{vmatrix}.$$

Observing that $H_1 = h_0 + r^{2w_3-2w_1}(h_1 + h_2)$ and $A_1{}^2 = a_0 + a_1 + r^{2w_3-2w_1}a_2$, we have

$$\kappa_1 = \frac{L_0N_0 - M_0{}^2}{A_1H_1}\left(1 + \frac{A_1{}^2(L_0N_0 - M_0{}^2)}{H_1{}^2} + \cdots\right)$$

$$= -\frac{w_3(w_3 - w_1)}{w_1\gamma_1(a_0 + a_1)^{1/2}}\begin{vmatrix} \gamma_2 & \gamma_2' \\ \gamma_3 & \gamma_3' \end{vmatrix}(1 + O(r^{2w_3-2w_1})).$$

We also conclude that

$$\kappa_2 = \frac{h_0}{(a_0 + a_1)^{3/2}}r^{-w_3}(1 + O(r^2)),$$

$$\tilde{v}_1 = (N_0 - \tfrac{(\gamma_1')^2(L_0N_0-M_0{}^2)}{h_0} + O(r^2))\partial_r + O(r)\partial_s,$$

$$\tilde{v}_2 = (N_0 - \tfrac{(\gamma_1')^2 h_0}{w_1{}^2(\gamma_1\gamma_2'-\gamma_1'\gamma_2)^2} + O(r^2))r\partial_r$$
$$- (M_0 - \tfrac{\gamma_1\gamma_1' h_0}{w_1(\gamma_1\gamma_2'-\gamma_1'\gamma_2)^2} + O(r^2))\partial_s.$$

Finally we obtain that

$$\tilde{v}_1\kappa_1 = r^{w_3-2w_1-1}\Big[(w_3 - 2w_1)(N_0 - \tfrac{(\gamma_1')^2(L_0N_0-M_0{}^2)}{h_0})\tfrac{L_0N_0 - M_0{}^2}{(a_0+a_1)^{1/2}h_0}$$
$$+ O(r)\Big],$$

$$\tilde{v}_2\kappa_1 = r^{w_3-2w_1}\Big[(w_3 - 2w_1)(N_0 - \tfrac{(\gamma_1')^2 h_0}{w_1{}^2(\gamma_1\gamma_2'-\gamma_1'\gamma_2)^2})\frac{L_0N_0 - M_0{}^2}{(a_0 + a_1)^{1/2}h_0}$$
$$- (M_0 - \tfrac{\gamma_1\gamma_1' h_0}{w_1(\gamma_1\gamma_2' - \gamma_1'\gamma_2)^2})(\frac{L_0N_0 - M_0{}^2}{(a_0 + a_1)^{1/2}h_0})' + O(r)\Big],$$

$$\tilde{v}_1\kappa_2 = \left[(w_3 - 2w_1)(N_0 - \frac{(\gamma_1')^2(L_0N_0 - M_0{}^2)}{h_0})\frac{h_0}{(a_0 + a_1)^{3/2}} + O(r)\right],$$

$$\tilde{v}_2\kappa_2 = r^{-w_3}\left[-w_3(N_0 - \frac{(\gamma_1')^2 h_0}{w_1{}^2(\gamma_1\gamma_2' - \gamma_1'\gamma_2)^2}\frac{h_0}{(a_0+a_1)^{3/2}})\right.$$
$$\left. - (M_0 - \frac{\gamma_1\gamma_1' h_0}{w_1(\gamma_1\gamma_2' - \gamma_1'\gamma_2)^2})(\frac{h_0}{(a_0 + a_1)^{3/2}})' + O(r^2)\right].$$

3. Focal curves: Case $(w_1, w_2, w_3) = (1, 2, 2)$

Assume that $(w_1, w_2, w_3) = (1, 2, 2)$. Then we have

$$\mathbf{n}(0, s) = \frac{1}{|\mathbf{a}|}\mathbf{a} \quad \text{where} \quad \mathbf{a} = \left(0, -\begin{vmatrix} \gamma_1 & \gamma_1' \\ 2\gamma_3 & \gamma_3' \end{vmatrix}, \begin{vmatrix} \gamma_1 & \gamma_1' \\ 2\gamma_2 & \gamma_2' \end{vmatrix}\right).$$

We assume $\mathbf{a}$ is not 0 for all s. The tangent plane at the origin degenerates into a line or point. We call the plane orthogonal to such a line at the origin the **normal plane**. The focal set (or caustic) is given by $\Sigma = \{p \in \mathbf{R}^3 : p = g^t(S(g^t))\}$, where $S(g^t)$ is the set of singular points of g^t. Since $\mathbf{n}(0, s)$ lies in the normal plane, $\Sigma|_{r=0}$ also lies in the normal plane. The set $\Sigma|_{r=0}$ is the image of the map

$$\phi : S^1 \to \mathbf{R}^3, \quad s \mapsto \frac{1}{\kappa_1(0, s)}\mathbf{n}(0, s) = \frac{\gamma_1}{2\begin{vmatrix} \gamma_2 & \gamma_2' \\ \gamma_3 & \gamma_3' \end{vmatrix}}\left(0, \begin{vmatrix} \gamma_1 & \gamma_1' \\ 2\gamma_3 & \gamma_3' \end{vmatrix}, -\begin{vmatrix} \gamma_1 & \gamma_1' \\ 2\gamma_2 & \gamma_2' \end{vmatrix}\right),$$

which we call by the **focal curve**. Focal curves are analogous to focal conics of Whitney umbrella (cf. [3, Lemma 3.3]).

Proposition 3.1. *A point $\phi(s_0)$ of the focal curve, which is not the origin, is subparabolic relative to $\tilde{v}_2$ if and only if it is a critical point of $|\phi(s)|^2$. This condition is equivalent to one of the following conditions.*

(1) ϕ is regular at $s = s_0$ and there is a circle centered at the origin which tangents to the focal curve at $\phi(s_0)$.

(2) ϕ is singular at $s = s_0$.

Proof. We observe that

$$\frac{d}{ds}|\phi(s)|^2 = -\frac{2\gamma_1 PQ}{(\gamma_2\gamma_3' - \gamma_3\gamma_2')^3}, \qquad (\tilde{v}_2\kappa_1)(0, s) = \frac{2PQ}{\gamma_1{}^2(a_0 + a_1)^{3/2}}$$

where

$$P = 2\gamma_1'^2\begin{vmatrix} \gamma_2 & \gamma_2' \\ \gamma_3 & \gamma_3' \end{vmatrix} - \gamma_1\begin{vmatrix} \gamma_1 & \gamma_1' & \gamma_1'' \\ 2\gamma_2 & \gamma_2' & \gamma_2'' \\ 2\gamma_3 & \gamma_3' & \gamma_3'' \end{vmatrix} \quad \text{and} \quad Q = \begin{vmatrix} 0 & \gamma_1 & \gamma_1' \\ -\gamma_3 & 2\gamma_2 & \gamma_2' \\ \gamma_2 & 2\gamma_3 & \gamma_3' \end{vmatrix}.$$

Since

$$\phi' = \frac{P}{(\gamma_2\gamma_3' - \gamma_3\gamma_2')^2}(0, \gamma_3, -\gamma_2),$$

$Q(s_0) = 0$ if and only if there is a circle centered at the origin which tangents to the focal curve at $\phi(s_0)$ whenever $\phi'(s_0) \neq 0$. We complete the proof. $\square$

Example 3.2. Here we show some pictures of focal curves for several $\gamma(s)$'s.

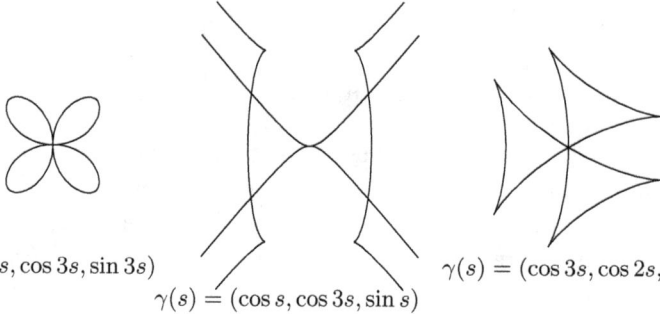

$\gamma(s) = (\cos 2s, \cos 3s, \sin 3s)$

$\gamma(s) = (\cos s, \cos 3s, \sin s)$

$\gamma(s) = (\cos 3s, \cos 2s, \sin 2s)$

4. Examples

Example 4.1. Let γ be a curve defined by $\gamma(s) = (\cos 2s, \sin 2s, \cos 3s)$. Setting $(w_1, w_2, w_3) = (1, 1, 2)$, we consider the weighted cone:

$$g(r, s) = (r\cos 2s, \ r\sin 2s, \ r^2\cos 3s).$$

Then we obtain $K|_{r=0} = \frac{1}{128}(7\cos 2s - 11)$, and $H|_{r=0} = (-1/4)\cos 3s$. So both principal curvatures are bounded near the locus defined by $r = 0$. The image of g is the same as the image of the map

$$\mathbf{R}^2 \to \mathbf{R}^3, \ (u, v) \mapsto (u^2 - v^2, \ 2uv, \ \sqrt{u^2 + v^2}(u^3 - 3uv^2)).$$

Let $g_i(u, v)$ $(i = 1, 2, 3)$ be a homogeneous polynomial of degree w_i and set $g(u, v) = (g_1(u, v), g_2(u, v), g_3(u, v))$. Considering $\gamma(s)$ a parametrization of a connected component of $S^2 \cap g(\mathbf{R}^2)$, we obtain many examples of weighted cones.

Example 4.2. Consider the curve defined by

$$\gamma(s) = (\cos s, \cos s \sin s, \sin^2 s).$$

The map $g(r, s) = (r\cos s, r^2 \cos s \sin s, r^2 \sin^2 s)$ is a weighted cone with $(w_1, w_2, w_3) = (1, 2, 2)$ and its image is known as Whitney umbrella. Remark that the focal curve of g coincides with the focal conic of the standard

Whitney umbrella defined by $(u, v) \mapsto (u, uv, v^2)$. We then have

$$E = \cos^2 s + 4r^2 \sin^2 s,$$
$$F = r \sin s \cos s(-1 + 2r^2),$$
$$G = r^2 \sin^2 s + r^4,$$
$$EG - F^2 = r^4(\cos^2 s + 4 \sin^2 s) + 4r^6 \sin^4 s,$$
$$\mathbf{n} = \frac{1}{\sqrt{\cos^2 s + 4 \sin^2 s + 4r^2 \sin^4 s}}(2r \sin^2 s, 2 \sin s, \cos s),$$
$$L = -\frac{2 \sin^2 s \cos s}{\sqrt{\cos^2 s + 4 \sin^2 s + 4r^2 \sin^4 s}},$$
$$M = \frac{2r \sin^3 s}{\sqrt{\cos^2 s + 4 \sin^2 s + 4r^2 \sin^4 s}},$$
$$N = \frac{2r^2 \cos s(1 + \sin^2 s)}{\sqrt{\cos^2 s + 4 \sin^2 s + 4r^2 \sin^4 s}},$$
$$K = -\frac{4 \sin^2 s}{r^2(\cos^2 s + 4 \sin^2 s + r^2 \sin^4 s)^2},$$
$$H = \frac{\cos s(1 + 3r^2 \sin^2 s)}{r^2(\cos^2 s + 4 \sin^2 s + 4r^2 \sin^4 s)^{\frac{3}{2}}}.$$

We have

$$\kappa_1 = -\frac{2 \sin s \tan s}{\sqrt{\cos^2 s + 4 \sin^2 s}} + \frac{4r^2 \sin^4 s(2 \cos s + \sin s \tan s(11 + 8 \tan^2 s))}{(\cos^2 s + 4 \sin^2 s)^{\frac{3}{2}}}$$
$$- \frac{4r^4 \sin^6 s(10 \cos^3 s + 108 \cos s \sin^2 s + \sin^3 s \tan s(391 + 528 \tan^2 s + 256 \tan^4 s))}{(\cos^2 s + 4 \sin^2 s)^{\frac{5}{2}}}$$
$$+ O(r^6),$$

$$\kappa_2 = \frac{1}{r^2}\left(\frac{2 \cos s}{\sqrt{\cos^2 s + 4 \sin^2 s}} + \frac{4r^2(2 \cos^3 s \sin^2 s + 7 \cos s \sin^4 s + 8 \sin^5 s \tan s)}{(\cos^2 s + 4 \sin^2 s)^{\frac{3}{2}}} + O(r^4)\right).$$

Observe cancellation of the coefficients of r^0, r^2 of $L - \kappa_1 E$ and the coefficient of r of $M - \kappa_1 F$:

$$L - \kappa_1 E = \frac{(33 - 28 \cos 2s + 3 \cos 4s)^2 \sin^3 s \tan^3 s}{(\cos^2 s + 4 \sin^2 s)^{\frac{3}{2}}} r^4 + O(r^6),$$
$$M - \kappa_1 F = \frac{(33 - 28 \cos 2s + 3 \cos 4s) \sin s \tan^2 s}{2(\cos^2 s + 4 \sin^2 s)^{\frac{3}{2}}} r^3 + O(r^5),$$
$$N - \kappa_1 G = \frac{2}{\cos s \sqrt{\cos^2 s + 4 \sin^2 s}} r^2 + O(r^4).$$

Moreover, we have

$$M - \kappa_2 F = \frac{1}{r}\left(\frac{2\cos^2 s \sin s}{(\cos^2 s + 4\sin^2 s)^{\frac{3}{2}}} + O(r^2)\right),$$

$$N - \kappa_2 G = -\frac{2\cos s \sin^2 s}{(\cos^2 s + 4\sin^2 s)^{\frac{3}{2}}} + O(r^2).$$

Then we have

$$\tilde{v}_1 = \left(\frac{2}{\cos s \sqrt{\cos^2 s + 4\sin^2 s}} + O(r^2)\right)\partial_r$$
$$+ \left(-\frac{(33 - 28\cos 2s + 3\cos 4s)\sin s \tan^2 s}{2(\cos^2 s + 4\sin^2 s)^{\frac{3}{2}}}r + O(r^3)\right)\partial_s,$$

$$\tilde{v}_2 = \left(-\frac{2\cos s \sin^2 s}{(\cos^2 s + 4\sin^2 s)^{\frac{3}{2}}}r + O(r^3)\right)\partial_r$$
$$+ \left(-\frac{2\cos^2 s \sin s}{(\cos^2 s + 4\sin^2 s)^{\frac{3}{2}}} + O(r^2)\right)\partial_s.$$

We conclude that

$$\tilde{v}_1\kappa_1|_{r=0} = 0,$$

$$\tilde{v}_1^2\kappa_1|_{r=0} = \frac{96(-4 + 3\cos^2 s)^2 \tan^4 s}{\cos s(\cos^2 s + 4\sin^2 s)^{\frac{7}{2}}},$$

$$\tilde{v}_2\kappa_1|_{r=0} = \frac{8(2 - \cos^2 s)\sin^2 s}{(\cos^2 s + 4\sin^2 s)^3}.$$

Hence, from Theorem 2.5 it follows that if $\cos s_0 \neq 0$ or $\sin s_0 \neq 0$ then $g^{1/\kappa_1(0,s_0)}$ has a swallowtail singularity at $g^{1/\kappa_1(0,s_0)}(0, s_0)$.

References

1. V. I. Arnol'd, S. M. Gusein-Zade and A. N. Varchenko, Singularities of differentiable maps I, Birkhäuser, 1986.
2. T. Fukui and M. Hasegawa, Singularities of parallel surfaces, to appear in Tohoku Magthematical Journal **64** (2012).
3. T. Fukui and M. Hasegawa, Fronts of Whitney umbrella, Journal of Singularities **4** (2012), 35–67.
4. M. Kokubu, W. Rossman, K. Saji, M. Umehara and K. Yamada, Singularities of flat fronts in hyperbolic 3-space, Pacific J. of Math. **221** (2005), no. 2, 303–351.

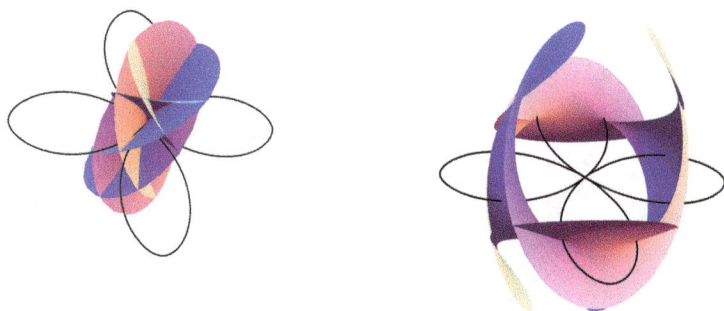

Fig. 1. Examples of a weighted cone, its focal curve, and its front: $g(r,s) = (r\cos 2s, r^2\cos 3s, r^2\sin 3s)$ (left), $g^t(r,s) = g(r,s) + \mathbf{n}(r,s)/3$ (right).

Fig. 2. Examples of a weighted cone and its front: $g(r,s) = (r\cos 2s, r\sin 2s, r^2\cos 3s)$ (left), $g^t(r,s) = g(r,s) + 3/2\mathbf{n}(r,s)$ (right).

Involutive deformations of the regular part of a normal surface

Adam Harris and Kimio Miyajima

Department of Mathematics and Statistics, School of Science and Technology,
University of New England Armidale, NSW 2351 Australia
adamh@turing.une.edu.au

Department of Mathematics and Computer Science, Faculty of Science,
Kagoshima University, Kagoshima 890, Japan
miyajima@sci.kagoshima-u.ac.jp

We define the property of *involutivity* for deformations of complex structure on a manifold X, with particular reference to the regular part of a normal surface. Our main result is a sufficient condition for involutivity in terms of a "$\bar{\partial}$-Cartan formula", previously examined in [3] in the more special context of cone singularities. By way of examples we show that some involutive deformations of the regular part determine a subspace, if not the entire versal space, of flat deformations of normal surface singularities, while others may determine Stein surfaces which lie outside the versal space of flat deformations of a given normal surface.

Keywords: normal singularity, complex deformation, Stein completion
AMS classification numbers: 32S30, 32G05, 32C15

1. Introduction

A $\bar{\partial}$-closed form $\psi \in C^\infty(X, T_X^{1,0} \otimes (T_X^{0,1})^*)$ on any complex manifold X will be said to represent an *involutive* deformation of the complex structure if the Frölicher-Nijenhuis bracket $[\psi, \psi]_{FN}$ vanishes identically. Such deformations automatically satisfy the Kodaira-Spencer integrability equation independently of their analytic properties, and are therefore well-suited to the context of non-compact spaces, provided they can be shown to represent a significant subspace of the tangent cohomology $H^1(X, T_X)$. It was shown in [3] that this subspace is non-trivial for $X = V \setminus \{0\}$, where $V \subseteq \mathbb{C}^N$ is a complex variety with isolated cone-singularity at the origin, and is in fact infinite-dimensional when V is a surface of this type. Given a potentially large space of integrable complex deformations of $V \setminus \{0\}$, the following should be noted:

- On the "link" defined by $V \cap S^{2N-1}$ each complex deformation induces a corresponding deformation of the CR-structure. If $dim_{\mathbb{C}}(V) \geq 3$, and if strict pseudoconvexity of the initial CR-structure is maintained by a given deformation, then a well-known theorem of Rossi [5] ensures that the ambient complex structure may be completed to a normal Stein space V' which is biholomorphically distinct from V.
- There is no a priori reason to expect that the Stein embedding dimension of V', i.e., $V' \subset \mathbb{C}^{N'}$ should be such that $N' = N$. In fact, it was shown in [4] that only *stably embeddable* CR-deformations of the link of an isolated singularity are in one-to-one correspondence with the finite-dimensional versal space of flat algebraic deformations of V.

In light of these remarks, the question to be addressed in this article is whether there exist non-trivial deformations $[\psi] \in H^1(V \setminus \{0\}, T_V)$ such that

- $[\psi]$ is involutive
- there exists a normal Stein completion V' of the associated deformation of $V \setminus \{0\}$
- V' does not belong to the versal space of flat deformations of V.

When $dim_{\mathbb{C}}(V) = 2$ the existence of a Stein completion is guaranteed if and only if the induced strongly pseudoconvex CR-structure of an integrable deformation (involutive or otherwise) is embeddable in some complex Euclidean space [6]. Equivalently, it has been shown in [1] that a two-dimensional *1-corona* Y (cf. section 3) admits a Stein compactification (unique after normalization) if and only if the Cauchy-Riemann operator

$$\bar{\partial}_Y : C^{\infty}(Y, \mathbb{C}) \to C^{\infty}(Y, (T_Y^{0,1})^*)$$

has closed range with respect to the C^{∞}-topology.

Our first result, in section 2, treats the existence of involutive deformations on a complex surface.

Theorem: *Let X be a smooth complex surface, and $\sigma \in H^0(X, T_X)$ a holomorphic vector field. Let*

$$L_{\bar{\sigma}} : C^{\infty}(X, T_X^{1,0} \otimes (T_X^{0,1})^*) \to C^{\infty}(X, T_X^{1,0} \otimes (T_X^{0,1})^*)$$

represent the natural first-order operator defined by Lie differentiation along the anti-holomorphic vector field $\bar{\sigma}$. Then a sufficient condition for any class

$[\psi] \in H^1(X, T_X)$ to be an involutive deformation is that it belong to the quotient

$$\Gamma^1_{\bar{\sigma}} := \frac{im(L_{\bar{\sigma}}) \cap ker(\bar{\partial})}{im(\bar{\partial})} .$$

This is an easy generalization of the surface-case of cone-singularities, $X = V \setminus \{0\}$, as treated in [3]. We recall briefly that these singularities are obtained specifically by blowing down the zero-section of a negative hermitian-holomorphic line bundle defined over a Riemann surface. The infinite-dimensionality of $\Gamma^1_{\bar{\sigma}}$ was demonstrated with respect to a particular vector field $\bar{\sigma}$, the intrinsic definition of which is reviewed in section 3.

In the final section we present two very different examples in which involutive deformation of the regular part of an isolated surface singularity can admit a Stein completion. The first follows the approach to constructing the versal space of flat deformations of an isolated singularity developed in [4], and is applied specifically to the case of the orbifold surface singularity of multiplicity two. The second example is of a complex-analytic family of involutive deformations, each fibre of which blows-down to a normal Stein surface in such a way that they are *not* stably embedded in complex Euclidean space. It is easily derived from a construction of unstable perturbations of an embedded CR-structure of dimension three, due to Catlin and Lempert [2]. From this we conclude that the versal space of flat deformations of an isolated singularity, although it may contain a subspace of involutive deformations of the regular part, does not in general contain all Stein-complete involutive deformations.

2. Involutive deformations of surfaces

At first, let X be an arbitrary complex manifold, with $\varphi \in C^\infty(X, T_X^{1,0} \otimes \bigwedge^q(T_X^{0,1})^*)$ written locally in the form

$$\varphi = \Sigma_{(\lambda)} \Sigma_{1 \leq \alpha \leq n+1} \varphi^\alpha_{(\lambda)} \partial_\alpha \otimes d\bar{w}_{(\lambda)} .$$

Here $T_X^{1,0}$, as usual, denotes the holomorphic tangent bundle, with local basis $\{\partial_\alpha\}_{\alpha=1}^n$, (λ) is a multi-index $(\lambda_{i_1}, ..., \lambda_{i_q})$ and $\varphi^\alpha_{(\lambda)}$ is smooth in any local chart. The following calculation, reproduced for the reader's convenience from [3], is carried out for $q = 1$, though the general case is essentially well-known. A vector field $\xi^{0,1}$ will be referred to as "anti-holomorphic" if the Lie bracket $[\xi^{0,1}, \vartheta] = 0$ for any holomorphic vector field ϑ, and will be written locally in the form

$$\xi^{0,1} = \Sigma_{\nu=1}^n \xi^\nu \bar{\partial}_\nu .$$

Now consider the contraction

$$\iota_{\xi^{0,1}}\varphi^\alpha = \Sigma_{\lambda=1}^n \varphi_\lambda^\alpha \xi^\lambda \quad \text{such that}$$

$$\left(\bar\partial \iota_{\xi^{0,1}}\varphi^\alpha\right)_\nu = \Sigma_\lambda \left(\frac{\partial \varphi_\lambda^\alpha}{\partial \bar w_\nu}\xi^\lambda + \varphi_\lambda^\alpha \frac{\partial \xi^\lambda}{\partial \bar w_\nu}\right).$$

On the other hand,

$$(\bar\partial\varphi)_{\lambda,\nu} = \Sigma_{\lambda<\nu}\left(\frac{\partial\varphi_\nu^\alpha}{\partial\bar w_\lambda} - \frac{\partial\varphi_\lambda^\alpha}{\partial\bar w_\nu}\right)$$

so that

$$\left(\iota_{\xi^{0,1}}\bar\partial\varphi\right)_\nu = \Sigma_{\lambda<\nu}\xi^\lambda\left(\frac{\partial\varphi_\nu^\alpha}{\partial\bar w_\lambda} - \frac{\partial\varphi_\lambda^\alpha}{\partial\bar w_\nu}\right) - \Sigma_{\lambda>\nu}\xi^\lambda\left(\frac{\partial\varphi_\lambda^\alpha}{\partial\bar w_\nu} - \frac{\partial\varphi_\nu^\alpha}{\partial\bar w_\lambda}\right)$$

$$= \Sigma_{\lambda\neq\nu}\xi^\lambda\frac{\partial\varphi_\nu^\alpha}{\partial\bar w_\lambda} - \Sigma_{\lambda\neq\nu}\xi^\lambda\frac{\partial\varphi_\lambda^\alpha}{\partial\bar w_\nu},$$

and therefore

$$\left(\bar\partial\iota_{\xi^{0,1}}\varphi^\alpha\right)_\nu + \left(\iota_{\xi^{0,1}}\bar\partial\varphi^\alpha\right)_\nu = \frac{\partial\varphi_\nu^\alpha}{\partial\bar w_\nu}\xi^\nu + \Sigma_\lambda\,\varphi_\lambda^\alpha\frac{\partial\xi^\lambda}{\partial\bar w_\nu} + \Sigma_{\lambda\neq\nu}\xi^\nu\frac{\partial\varphi_\nu^\alpha}{\partial\bar w_\lambda}$$

$$= \xi^{0,1}(\varphi_\nu^\alpha) + \Sigma_\lambda\,\varphi_\lambda^\alpha\frac{\partial\xi^\lambda}{\partial\bar w_\nu}.$$

Writing $\varphi_\nu = \Sigma_\alpha\varphi_\nu^\alpha\partial_\alpha$, we note that the assumption $\xi^{0,1}$ is anti-holomorphic implies specifically that

$$[\xi^{0,1},\varphi_\nu]^\alpha = \xi^{0,1}(\varphi_\nu^\alpha).$$

In addition,

$$[\xi^{0,1},\bar\partial_\nu] = -\Sigma_\lambda\frac{\partial\xi^\lambda}{\partial\bar w_\nu}\bar\partial_\lambda,$$

so that

$$\Sigma_\lambda\,\varphi_\lambda^\alpha\frac{\partial\xi^\lambda}{\partial\bar w_\nu} = -\varphi^\alpha([\xi^{0,1},\bar\partial_\nu])$$

implies

$$\left(\bar\partial\iota_{\xi^{0,1}}\varphi^\alpha\right)_\nu + \left(\iota_{\xi^{0,1}}\bar\partial\varphi^\alpha\right)_\nu = \xi^{0,1}(\varphi_\nu^\alpha) - \varphi^\alpha([\xi^{0,1},\bar\partial_\nu]) = \left(\mathcal{L}_{\xi^{0,1}}\varphi^\alpha\right)_\nu,$$

which recalls the famous formula for the Lie derivative due to E. Cartan.

In particular, let $\sigma \in H^0(X, T_X^{1,0})$ be a holomorphic vector field on X, and consider

$$\psi \in C^\infty(X, T_X^{1,0} \otimes (T_X^{0,1})^*)$$

such that $\bar{\partial}\psi = 0$ and $\psi = L_{\bar{\sigma}}\varphi$, for some smooth $\varphi \in C^\infty(X, T_X^{1,0} \otimes (T_X^{0,1})^*)$. It follows from the formula above that ψ is cohomologous to $\iota_{\bar{\sigma}}\bar{\partial}\varphi$.

Writing ψ in local form as $\Sigma_{\alpha,\lambda}\psi_\lambda^\alpha \partial_\alpha \otimes d\bar{w}_\lambda$ we recall the specific formula for the Frölicher-Nijenhuis bracket

$$[\psi, \psi]_{FN} = \Sigma_{\alpha,\beta=1}^n 2\left(\psi^\alpha \wedge \partial_\alpha \psi^\beta\right)\partial_\beta = \Sigma_{1 \le \lambda < \nu \le n}[\psi_\lambda, \psi_\nu]d\bar{w}_\lambda \wedge d\bar{w}_\nu ,$$

where $[\psi_\lambda, \psi_\nu]$ denotes the standard Lie bracket of vector fields. Let $\eta \in C^\infty(X, T_X^{1,0} \otimes \bigwedge^2(T_X^{0,1})^*)$ be represented locally as $\eta_{\alpha,\beta}^\gamma \partial_\gamma \otimes d\bar{w}_\alpha \wedge d\bar{w}_\beta$. Now

$$(\iota_{\bar{\sigma}}\eta)_\lambda^\gamma = \Sigma_{\alpha<\lambda}\bar{\sigma}^\alpha \eta_{\alpha,\lambda}^\gamma - \Sigma_{\beta>\lambda}\bar{\sigma}^\beta \eta_{\lambda,\beta}^\gamma ,$$

and in particular, $n = 2$ implies

$$(\iota_{\bar{\sigma}}\eta)_1^\gamma = -\bar{\sigma}^2 \eta_{12}^\gamma , \quad (\iota_{\bar{\sigma}}\eta)_2^\gamma = \bar{\sigma}^1 \eta_{12}^\gamma ,$$

so that

$$[\iota_{\bar{\sigma}}\eta, \iota_{\bar{\sigma}}\eta]_{FN} = (-\bar{\sigma}^2 \eta_{12}(\bar{\sigma}^1) + \bar{\sigma}^1 \eta_{12}(\bar{\sigma}^2))\eta_{12} = 0 ,$$

given $\bar{\sigma}$ is anti-holomorphic.

We summarize with the following

Theorem 2.1. *Let X be a smooth complex surface, and $\sigma \in H^0(X, T_X)$ a holomorphic vector field. Let*

$$L_{\bar{\sigma}} : C^\infty(X, T_X^{1,0} \otimes (T_X^{0,1})^*) \to C^\infty(X, T_X^{1,0} \otimes (T_X^{0,1})^*)$$

represent the natural first-order operator defined by Lie differentiation along the anti-holomorphic vector field $\bar{\sigma}$. Then a sufficient condition for any class $[\psi] \in H^1(X, T_X)$ to be an involutive deformation is that it belong to the quotient

$$\Gamma_{\bar{\sigma}}^1 := \frac{im(L_{\bar{\sigma}}) \cap ker(\bar{\partial})}{im(\bar{\partial})} .$$

Proof. If $\psi = L_{\bar{\sigma}}\varphi$ then $[\psi] \in H^1(X, T_X)$ is also represented by $\iota_{\bar{\sigma}}\bar{\partial}\varphi$, which is clearly involutive when $n = 2$. $\qquad\square$

3. Some remarks on Stein completion

Our primary interest is in deformations of the regular part of a normal complex surface, in particular $X = V \setminus \{0\}$, where $V \subset \mathbb{C}^N$ is a surface with isolated singularity at the origin. When V corresponds to the neighbourhood of a "cone" singularity, i.e., obtained by blowing-down the zero section of a negative holomorphic line bundle over a Riemann surface, then it was shown in [3] that $\Gamma^1_{\bar\sigma}$ is in fact infinite-dimensional as a complex vector space. The specific vector field $\bar\sigma$ used in [3] was defined intrinsically as the Kähler metric-dual of $\partial\rho$, where ρ is a strongly plurisubharmonic function on the complement of the zero section, defined relative to the hermitian metric of the line bundle in the standard way. In general small deformations, when restricted to the link $\Sigma := V \cap \mathbb{S}^{2N-1} \subset \mathbb{C}^N$ induce distinct CR-structures which preserve the strict pseudoconvexity of the initial structure. If ρ is a strongly plurisubharmonic function on $X = V \setminus \{0\}$, such that $\rho(0) = 0$ and $\Sigma = \rho^{-1}(1)$, then the same conclusion can be drawn simultaneously for all level sets $\rho = \varepsilon$, $0 < \varepsilon_0 \leq \varepsilon \leq 1$. A small involutive deformation ψ therefore induces a smooth function ρ' on a surface X' which is strongly plurisubharmonic on (the "1-corona") $Y_{\varepsilon_0} = \{\varepsilon_0 \leq \rho' \leq 1\}$. Let $\bar\sigma^\psi$ denote a vector field on Y_{ε_0} which is anti-holomorphic with respect to the complex structure induced by an involutive ψ, i.e.,

$$\bar\sigma^\psi = \bar\sigma + \psi(\bar\sigma) ,$$

where $\bar\sigma$ is anti-holomorphic with respect to the initial complex structure on X. Hence, for $u \in C^\infty_{\mathbb{C}}(Y_{\varepsilon_0})$, we have

$$\bar\partial^\psi u(\bar\sigma^\psi) = du(\bar\sigma^\psi) = \bar\partial u(\bar\sigma) + \partial u(\psi(\bar\sigma)) .$$

Theorem 1.1 of [1] states that for the existence of a Stein compactification of Y_{ε_0} it is necessary and sufficient that $\bar\partial^\psi$ have *closed range* relative to the C^∞-topology on $C^\infty(Y_{\varepsilon_0}, (T^{0,1}_{Y_{\varepsilon_0}})^*)$.

In the following examples we present two very different ways in which involutive deformation of the regular part of an isolated surface singularity can admit a Stein completion. The first follows the approach to constructing the versal space of flat deformations of an isolated singularity developed in [4]. It is applied specifically to the case of the orbifold surface singularity of multiplicity two as follows.

Let $X := \mathbb{C}^2 \setminus \{(0,0)\}$ with coordinate (z,w), and denote $r := |z|^2 + |w|^2$. Fix a global frame $\{Z, \xi\}$ of $T^{1,0}X$ such that

$$Z := \bar{w}\partial_z - \bar{z}\partial_w , \text{ and } \xi := z\partial_z + w\partial_w .$$

(While the above is a convenient definition for ξ, it plays no specific role in the following calculation.) Now define

$$\psi := \frac{1}{r^2} Z \otimes (\bar{Z})^* \in C^\infty(X, T^{1,0}_X \otimes (T^{0,1}_X)^*) .$$

Note that $\bar\partial\psi = 0$ and $[\psi, \psi]_{FN} = 0$ since, in the first instance,

$$\bar\partial\psi(\bar{Z}, \bar\xi) = [\bar{Z}, \psi(\bar\xi)]_{T^{1,0}X} - [\bar\xi, \psi(\bar{Z})]_{T^{1,0}X} - \psi([\bar{Z}, \bar\xi]) ,$$

where the subscript $T^{1,0}X$ implies taking the $(1,0)$-part of the Lie bracket, and by a direct calculation this is seen to vanish. On the other hand, it is clear that

$$[\psi, \psi]_{FN}(\bar{Z}, \bar{\xi}) = [\psi(\bar{Z}), \psi(\bar{\xi})] = 0 \ .$$

Next consider a $\mathbb{Z}_2$-action $(z, w) \mapsto (-z, -w)$ on X. Since ψ is invariant under this $\mathbb{Z}_2$-action, it is considered as a form on X_1, say $\psi_1 \in C^\infty(X_1, T^{1,0}_{X_1} \otimes (T^{0,1}_{X_1})^*)$, where $X_1 := X/\mathbb{Z}_2$. $(X_1, T^{0,1}X_1)$ is realized as the regular part of a Stein space V_1 defined by an equation $\zeta_1\zeta_3 - \zeta_2^2 = 0$ in $\mathbb{C}^3$, with embedding $X_1 \ni (z, w) \mapsto (\zeta_1, \zeta_2, \zeta_3) = (z^2, zw, w^2) \in \mathbb{C}^3$.

Now $t\psi_1$ defines an involutive deformation of complex structure on X_1, and by a specific application of the theory of [4] it is possible to compute a family of embeddings

$$X_1 \ni (z, w) \mapsto (z^2 - t\frac{\bar{w}^2}{r^2}, zw + t\frac{\bar{z}\bar{w}}{r^2}, w^2 - t\frac{\bar{z}^2}{r^2}) \in \mathbb{C}^3$$

which realizes $(X_1, {}^{t\psi_1}T^{0,1}X_1)$ as a stable family of submanifolds of $\mathbb{C}^3$. It follows that $(X_1, {}^{t\psi_1}T^{0,1}X_1)$ admits a simultaneous family of Stein completions in $\mathbb{C}^3$ defined by

$$\zeta_1\zeta_3 - \zeta_3^2 = -t.$$

This is clearly recognizable as the versal family of flat deformations of the orbifold singularity.

By contrast, we now present an example of an involutive deformation of the regular part of a normal surface which is *not* stably embedded, based on the construction of unstable CR-perturbations introduced in [2]. While existence of a Stein completion for each individual member of this family is guaranteed, it is not simultaneously so, as will be seen below.

Let C be a compact Riemann surface of genus g, and $Pic_k(C)$ one of the connected components, i.e., a coset of the Jacobi variety, inside $Pic(C)$. A natural holomorphic family of ruled surfaces $\pi : \mathcal{F} \to Pic_k(C)$ is defined, such that for all $\lambda \in Pic_k(C)$, $\pi^{-1}(\lambda)$ corresponds to the total space of the associated line bundle of degree k over C. If $\Lambda \to C \times Pic_k(C)$ is the relative line bundle of this family, it is clear that the direct image with respect to projection onto $Pic_k(C)$ of the sheaf of sections of Λ is not locally free - a fact which lies at the heart of the following construction. Following [2], we assume C contains a Weierstrass point p_0 such that the line bundle $\lambda_0 := [kp_0]$ is very ample for an appropriate k. A suitable example of C is given by the locus of

$$Z_0^k = Z_1^k + Z_2^k \text{ inside } \mathbb{P}_2 \ ,$$

for $k \geq 5$. Note that the genus is of course related to k by the formula

$$g = \frac{1}{2}(k-1)(k-2) \ .$$

Let $\Delta \subset \mathbb{C}$ be a small disc, and $\varphi : \Delta \to C$ a holomorphic map such that $\varphi(0) = p_0$. For $k \geq 5$ the line bundle λ_0 is very ample (in fact $h^0(C, \lambda_0) = 3$), while neighbouring bundles of the form $\lambda_t := [k\varphi(t)]$ are *not* very ample (in fact $h^0(C, \lambda_t) = 1$, $t \neq 0$). Now consider the (holomorphic) map

$$\psi : \Delta \to Pic_k(C) \ , \ \psi(t) = [k\varphi(t)] \ .$$

If $\varpi : \mathcal{F}^* \to Pic_k(C)$ denotes the "dual family", i.e., each fibre $\varpi^{-1}(\lambda)$ denotes the total space of the dual line bundle λ^{-1}, then let $\psi^* \mathcal{F}^*$ denote its pullback over the disc. The fibres of this restricted family will "blow-down", via collapsing of the zero-section, to normal Stein surfaces provided each zero-section has a system of relatively compact, strictly pseudoconvex neighbourhoods. From the very-ampleness of λ_0 it is clear that a holomorphic map which collapses the zero-section of $\varpi^{-1}(\lambda_0)$ also embeds the resulting normal surface in $\mathbb{C}^3$. Hence there is a smooth function on $\varpi^{-1}(\lambda_0)$ which is strongly plurisubharmonic on the complement of its zero-section, and by the previous consideration of strictly pseudoconvex coronas being preserved under small deformations of the regular part of a singular surface, we arrive at an individual-fibre-wise reduction of $\psi^* \mathcal{F}^*$ to normal Stein surfaces. Since the natural embedding of the central fibre does not extend stably to its neighbours the assumption that such a blowing-down of members of this specific family is "simultaneous", i.e., that we have thereby defined an analytically fibred space, is unwarranted. The required condition for simultaneous blow-down is known to be the t-independence of $\dim H^1(X_t, \mathcal{O}_{X_t})$, which assures the extendability to neighbouring fibres of all holomorphic functions on the central fibre. Since all holomorphic functions on an analytic space are restrictions of holomorphic functions on an ambient $\mathbb{C}^N$, such extendability is equivalent to the stability of holomorphic embedding into $\mathbb{C}^N$.

It remains to see that the family of deformations of the central fibre is moreover involutive for all t. Given

$$Pic_k(C) \cong \mathbb{C}^g / \mathbb{Z}^{2g}$$

it follows that a small neighbourhood of $p_0 \in Pic_k(C)$ may be identified with any one of its translates via the integer lattice in $H^1(C, \mathcal{O}_C)$. Letting $L_0 := \varpi^{-1}(\lambda_0)$, and E_C denote the divisor corresponding to the embedded image of C as the zero-section of L_0, we recall the natural exact sequence of sheaves

$$0 \to \mathcal{O}_C \to \mathcal{T}_{L_0}(-\log(E_C))\,|_C \to \mathcal{T}_C \to 0 \ ,$$

from which it follows there is a natural injection $H^1(C, \mathcal{O}_C) \hookrightarrow \Gamma^1_{\bar\sigma}$ (cf. [3]). In this way, $\psi(t)$ is seen to be involutive for all $t \in \Delta$.

References

1. M. Brumberg and J. Leiterer, *On the Compactification of Concave Ends.* Math. Ann. **347** (2010), pp. 235-244.
2. D. Catlin and L. Lempert, *A Note on the Instability of Embeddings of Cauchy-Riemann Manifolds,* J. Geom. Anal. **2**, No. 2 (1992) pp. 99-104.
3. A. Harris and M. Kolar, *On Infinitesimal Deformations of the Regular Part of a Complex Cone Singularity,* Kyushu J. Math. **65**, No. 1 (2011), pp. 25-38.
4. K. Miyajima, *CR Construction of the Flat Deformations of Normal Isolated Singularities,* J. Alg. Geom. **8** (1999), pp. 403-470.
5. H. Rossi, *Attaching Analytic Spaces to an Analytic Space along a Pseudoconcave Boundary,* Proc. Conf. on Complex Analysis (Minneapolis, 1964), Springer, Berlin (1965), pp. 242-256.
6. S. S. T. Yau : *Kohn-Rossi Cohomology and its application to the Complex Plateau Problem.* Ann. Math. **113**, No. 1 (1981), pp. 67-110.

Connected components of regular fibers of differentiable maps

Dedicated to Professors Satoshi Koike and Laurentiu Paunescu on the occasion of their sixtieth birthdays

Jorge T. Hiratuka and Osamu Saeki

Departamento de Matemática, Instituto de Matemática e Estatística, Universidade de São Paulo, Caixa Postal 66.281, CEP: 05389-970, São Paulo, SP, Brazil
jotahira@ime.usp.br

Institute of Mathematics for Industry, Kyushu University, Motooka 744, Nishi-ku, Fukuoka 819-0395, Japan
saeki@imi.kyushu-u.ac.jp

For a map between smooth manifolds, the space of the connected components of its fibers is called the Stein factorization. In our previous paper, we showed that for generic smooth maps, the Stein factorizations are triangulable. As an application, we show that every connected component of a regular fiber is null-cobordant if the top dimensional homology of the Stein factorization vanishes.

Keywords: cobordism, regular fiber component, Stein factorization, generic map, triangulation
AMS classification numbers: 57R45, 57R75, 57R20, 57R05

1. Introduction

Let $f\colon M \to N$ be a generic C^∞ map between smooth manifolds. The space of the connected components of fibers of f is denoted by W_f. Then, we have the canonical quotient map $q_f\colon M \to W_f$ and the natural map $\bar{f}\colon W_f \to N$ such that $f = \bar{f} \circ q_f$. Such a decomposition of f into the composition of q_f and $\bar{f}$ is called the *Stein factorization* of f. Sometimes the quotient space W_f is also called the Stein factorization of f.

It is known that when $\dim M > \dim N$, the Stein factorization of $f\colon M \to N$, or the quotient space W_f, is a very important tool in studying the topological properties of the map f. Refer to [1,7–11,15], for example. In our previous paper [5], the authors have shown that the Stein factorization, and in particular the quotient space W_f, is triangulable for a large class of generic smooth maps f.

In this paper, we use the triangulation of the Stein factorization in order to study the cobordism classes of the components of regular fibers of generic smooth maps. It is known that if the target manifold N of a smooth map $f: M \to N$ is connected, then the regular fibers of f are all cobordant. However, the components of regular fibers may not be cobordant to each other. We show that for a generic smooth map $f: M \to N$, we can associate a top dimensional homology class $\gamma_f \in H_n(W_f)$, $n = \dim N$, in such a way that if f has a regular fiber component that is not null-cobordant, then γ_f does not vanish, where the coefficient group is the cobordism group of manifolds of dimension $m - n$, $m = \dim M$.

The paper is organized as follows. In §2 we give a precise definition of the Stein factorization of a continuous map between topological spaces and its triangulation. We also recall the cobordism group of manifolds, and state our main theorem. In §3 we define the homology class γ_f and prove our main theorem. We also give some enlightening examples. In §4, we show that the above homology class γ_f gives a bordism invariant for maps whose fibers are connected. We also give some observations which show that the cobordism classes of regular fibers have little relation to the cobordism class of the source manifold in general. Finally we give some related problems.

Throughout the paper, we will often abuse the terminology "simplicial complex" (or "simplicial map") to indicate the corresponding polyhedron (resp. PL map). The symbol "$\approx$" denotes a homeomorphism between topological spaces.

2. Preliminaries

In this section, we define the notion of a triangulation of the Stein factorization of a map and state our main theorem.

Definition 2.1. Let $g: X \to Y$ be a continuous map between topological spaces X and Y. Two points $x, x' \in X$ are g-*equivalent* if $g(x) = g(x')$ and the points x and x' are in the same connected component of $g^{-1}(g(x)) = g^{-1}(g(x'))$. We denote by W_g the quotient space with respect to the g-equivalence, endowed with the quotient topology. The quotient map is denoted by $q_g: X \to W_g$. Then there exists a unique continuous map $\bar{g}: W_g \to Y$ such that $g = \bar{g} \circ q_g$. The quotient space W_g or the commutative diagram

$$X \xrightarrow{\quad g \quad} Y$$

(with q_g going to W_g and $\bar{g}$ from W_g)

is called the *Stein factorization* of g.

There is a one-to-one correspondence between the quotient space and the space of the connected components of the fibers of g. Note that each fiber of the quotient map q_g is connected.

Remark 2.2. The space W_g is often called the *quotient space* or the *Reeb space* (or the *Reeb complex*) of g.

Let $g\colon X \to Y$ be a continuous map between topological spaces. Then, g is said to be *triangulable* if there exist simplicial complexes K and L, a simplicial map $s\colon K \to L$, and homeomorphisms $\lambda\colon |K| \to X$ and $\mu\colon |L| \to Y$ such that the following diagram is commutative:

$$
\begin{array}{ccc}
X & \xrightarrow{\ g\ } & Y \\
\lambda \uparrow & & \uparrow \mu \\
|K| & \xrightarrow{\ |s|\ } & |L|,
\end{array}
$$

where $|K|$ and $|L|$ are polyhedrons associated with K and L, respectively, and $|s|$ is the PL map associated with s.

In [5], the authors have proved the following.

Theorem 2.3. *Let $g\colon X \to Y$ be a proper continuous map between locally compact topological spaces X and Y. If g is triangulable, then so is the Stein factorization of g. More precisely, we have the commutative diagram* *for some finite simplicial complexes K', L' and V, some simplicial maps $s' \colon K' \to L'$, $\varphi \colon K' \to V$ and $\psi \colon V \to L'$, and some homeomorphisms λ, μ and Θ.*

Let us recall the notion of a cobordism of manifolds. Let M_0 and M_1 be closed oriented manifolds with $\dim M_0 = \dim M_1 (= m)$. We say that M_0 and M_1 are *oriented cobordant* if there exists a compact oriented $(m+1)$-dimensional manifold Q such that $\partial Q = (-M_0) \cup M_1$, where $-M_0$ denotes the manifold M_0 with the orientation reversed. Such a manifold Q is often called an *oriented cobordism* between M_0 and M_1. The above relation clearly defines an equivalence relation. The equivalence class of a manifold M will be denoted by $[M]$. We can define $[M] + [M'] = [M \cup M']$, in such a way that the set Ω_m of all oriented cobordism classes of closed oriented m-dimensional manifolds forms an additive group. This is called the *m-dimensional oriented cobordism group*.

In the above definition, if we ignore the orientations of the manifolds, then we get the *m-dimensional (unoriented) cobordism group*, denoted by $\mathfrak{N}_m$.

The groups Ω_m and $\mathfrak{N}_m$ have been extensively studied and their structures have been completely determined (see [16,17]). For example, the following is known.

- Ω_m is a finitely generated abelian group.
- $\mathfrak{N}_m$ is a finitely generated $\mathbb{Z}_2$-module.
- Ω_m is a finite group unless m is a multiple of four.
- $\Omega_0 \cong \mathbb{Z}$, $\Omega_1 = \Omega_2 = \Omega_3 = 0$, $\Omega_4 \cong \mathbb{Z}$, $\Omega_5 \cong \mathbb{Z}_2$, ...
- $\mathfrak{N}_0 \cong \mathbb{Z}_2$, $\mathfrak{N}_1 = 0$, $\mathfrak{N}_2 \cong \mathbb{Z}_2$, $\mathfrak{N}_3 = 0$, $\mathfrak{N}_4 \cong \mathbb{Z}_2^2$, $\mathfrak{N}_5 \cong \mathbb{Z}_2$, ...

A closed (oriented) manifold M with $[M] = 0$ is said to be *(oriented) null-cobordant*.

Our main theorem of this paper is the following.

Theorem 2.4. *Let M be a closed manifold and $f \colon M \to N$ a smooth map into a manifold N with $m = \dim M \geq \dim N = n$. Assume that f is triangulable (e.g. a topologically stable map). Then, we have the following.*

(1) If there exists a regular fiber component of f which is not null-cobordant, then $H_n(W_f; \mathbb{Z}_2) \neq 0$.

(2) Suppose that both M and N are oriented (note that then the regular fibers are naturally oriented). If there exists a regular fiber component of f which is not oriented null-cobordant, then $H_n(W_f; \Omega_{m-n}) \neq 0$.

3. Proof of Theorem 2.4

Proof. Let $s\colon K \to L$ be a triangulation of $f\colon M \to N$. Then, for the barycentric subdivision L' of L, there exist a subdivision K' of K and a simplicial map $s'\colon K' \to L'$ such that $|s'| = |s|$ (for example, see [6]). By Theorem 2.3 (see also [5]), we have a triangulation of the Stein factorization as in the commutative diagram

$$
\begin{array}{ccc}
M & \xrightarrow{\quad f \quad} & N \\
\end{array}
$$

where V is a finite simplicial complex of dimension n, $\varphi\colon K' \to V$ and $\psi\colon V \to L'$ are simplicial maps with ψ being non-degenerate, and λ, Θ and μ are homeomorphisms. Here, a simplicial map is *non-degenerate* if it preserves the dimension of each simplex.

For each n-simplex $\sigma \in V$, λ maps $|\varphi|^{-1}(b_\sigma)$ homeomorphically onto $q_f^{-1}(\Theta(b_\sigma))$, which is a component of a fiber of f, where b_σ is a point in the interior of σ. By the Sard theorem, we may assume that $\bar{f}(\Theta(b_\sigma))$ is a regular value of f. Then, define

$$
\omega_\sigma = [q_f^{-1}(\Theta(b_\sigma))] \in \mathfrak{N}_{m-n},
$$

which is the cobordism class of the regular fiber component corresponding to $\sigma \subset |V| \approx W_f$.

Lemma 3.1. *The cobordism class $[q_f^{-1}(\Theta(b_\sigma))]$ does not depend on a choice of $b_\sigma \in \operatorname{Int}\sigma$.*

Proof. Let b'_σ be another point in $\operatorname{Int}\sigma$ such that $\bar{f}(\Theta(b'_\sigma))$ is a regular value of f. We can choose an embedded arc γ in $\operatorname{Int}\sigma$ connecting b_σ and b'_σ in such a way that $\bar{f}(\Theta(\gamma))$ is transverse to f (for example, see [3, §4.3]). Then, $\lambda(|\varphi|^{-1}(\gamma))$ gives a smooth cobordism between $q_f^{-1}(\Theta(b_\sigma))$ and $q_f^{-1}(\Theta(b'_\sigma))$.

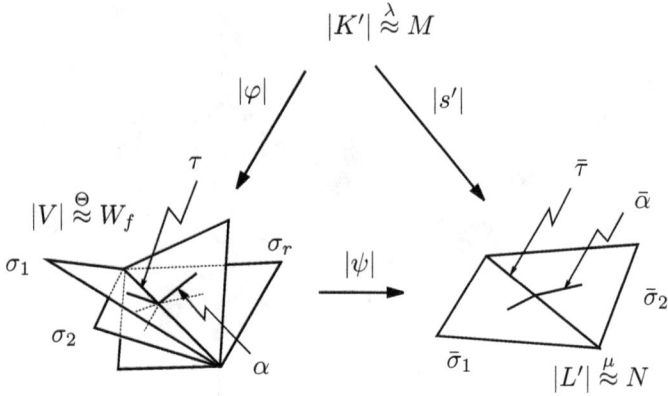

Fig. 1. The small arc $\bar{\alpha}$

$\square$

Set

$$c_f = \sum_\sigma \omega_\sigma \sigma \in C_n(V; \mathfrak{N}_{m-n}),$$

where σ runs over all n-simplices of V, and $C_n(V; \mathfrak{N}_{m-n})$ denotes the n-th chain group of V with coefficients in $\mathfrak{N}_{m-n}$.

Lemma 3.2. *We have $\partial c_f = 0$ in $C_{n-1}(V; \mathfrak{N}_{m-n})$, i.e. c_f is an n-cycle.*

Proof. Let τ be an arbitrary $(n-1)$-simplex of V, and let $\sigma_1, \sigma_2, \ldots, \sigma_r$ be the n-simplices of V containing τ as a face (see Fig. 1). We have only to show

$$\sum_{j=1}^r \omega_{\sigma_j} = 0$$

in $\mathfrak{N}_{m-n}$, i.e. the vanishing of the coefficient of τ in ∂c_f.

Let $\bar{\alpha}$ be a small arc in $|L'|$ which intersects $\bar{\tau} = \psi(\tau)$ transversely in one point (see Fig. 1), and let $\bar{\sigma}_1$ and $\bar{\sigma}_2$ be the n-simplices of L' adjacent to $\bar{\tau}$. Note that $\psi(\sigma_i)$ coincides with either $\bar{\sigma}_1$ or $\bar{\sigma}_2$. We take $\bar{\alpha}$ so that $\mu(\bar{\alpha})$ is a smooth arc in N transverse to f. Let α be the component of $|\psi|^{-1}(\bar{\alpha})$ that intersects τ. Then, $Q = q_f^{-1}(\Theta(\alpha))$ is an $(m-n+1)$-dimensional compact manifold and

$$\partial Q = \bigcup_{j=1}^r \lambda(|\varphi|^{-1}(b_{\sigma_j})).$$

Therefore, we have

$$\sum_{j=1}^{r} \omega_{\sigma_j} = \sum_{j=1}^{r} \left[\lambda(|\varphi|^{-1}(b_{\sigma_j})) \right] = 0$$

in $\mathfrak{N}_{m-n}$. □

Thus, c_f defines a homology class

$$\gamma_f \in H_n(W_f; \mathfrak{N}_{m-n}). \tag{3.1}$$

Furthermore, since $\dim W_f = n$, we have $\gamma_f \neq 0$ if and only if $c_f \neq 0$. Moreover, $c_f \neq 0$ if and only if there exists a component of a regular fiber of f which is not null-cobordant. Therefore, if such a regular fiber component exists, we have $H_n(W_f; \mathbb{Z}_2) \neq 0$, since $\mathfrak{N}_{m-n}$ is isomorphic to $\mathbb{Z}_2 \oplus \mathbb{Z}_2 \oplus \cdots \oplus \mathbb{Z}_2$, the direct sum of a finite number of copies of $\mathbb{Z}_2$.

The case where both M and N are oriented can be treated similarly. (In this case, the orientation of N induces an orientation of each n-simplex of V. Therefore, the argument also works with coefficients in the abelian group Ω_{m-n}.) This completes the proof of Theorem 2.4.

Example 3.1. Let us consider a tree T. Then, since $H_1(T; \mathbb{Z}_2) = 0$, there exists no Morse function $f_1 \colon M_1^5 \to \mathbb{R}$ on a closed 5-dimensional manifold M_1^5 such that the quotient space W_{f_1} is homeomorphic to T and that f_1 has $\mathbb{C}P^2$ as a component of a regular fiber. (Recall that $\mathbb{C}P^2$ is not null-cobordant.)

Example 3.2. There exists a Morse function $f_2 \colon M_2^5 \to \mathbb{R}$ on a closed 5-dimensional manifold M_2^5 whose quotient space is as depicted in Fig. 2. The integer at each vertex denotes the index of the corresponding critical point, and the 4-manifold attached to each edge denotes the corresponding regular fiber component.

Note that $H_1(W_{f_2}; \mathbb{Z}) \cong H_1(W_{f_2}; \Omega_4) \cong \mathbb{Z}$ is generated by the homology class γ_{f_2} of (3.1).

We note that the 5-dimensional manifold M_2^5 can be chosen to be diffeomorphic to $S^1 \times \mathbb{C}P^2$, which is null-cobordant.

Example 3.3. There exists a Morse function $f_3 \colon M_3^5 \to \mathbb{R}$ on a closed 5-dimensional manifold M_3^5 whose quotient space is as depicted in Fig. 3. Note that the quotient space W_{f_3} is homeomorphic to W_{f_2}; however, $\gamma_{f_3} = 0$ in $H_1(W_{f_3}; \mathbb{Z}) \cong \mathbb{Z}$, while $\gamma_{f_2} \neq 0$ in $H_1(W_{f_2}; \mathbb{Z})$. This means that even if the

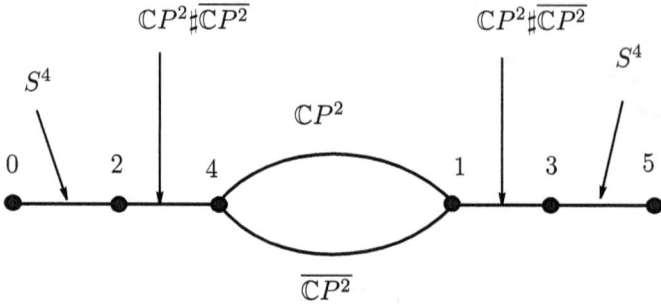

Fig. 2. An example with non-vanishing γ_{f_2}

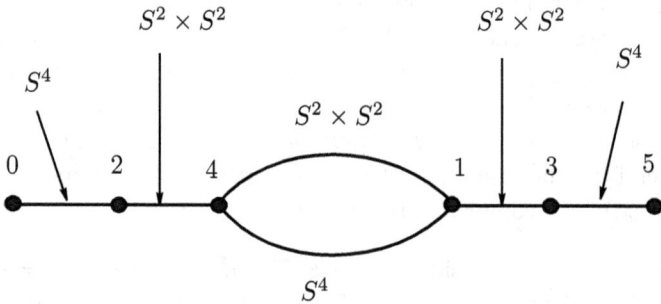

Fig. 3. An example with vanishing γ_{f_3}

top dimensional homology group of the quotient space does not vanish, the map may not have a regular fiber component that is not null-cobordant.

We note that the 5-dimensional manifold M_3^5 can be chosen to be diffeomorphic to $S^1 \times S^4$, which is null-cobordant.

By considering the product maps $\widetilde{f}_i = f_i \times \mathrm{id}_{S^k} : M_i^5 \times S^k \to \mathbb{R} \times S^k$, $k \geq 1$, $i = 1, 2, 3$, where id_{S^k} denotes the identity map of S^k, we can construct examples of higher dimensional quotient spaces as well.

Remark 3.1. For a smooth map $f : M \to N$ as in Theorem 2.4 with N being a closed manifold, we see that $\bar{f}_* \gamma_f \in H_n(N; \mathfrak{N}_{m-n}) \cong H_n(N; \mathbb{Z}_2) \otimes \mathfrak{N}_{m-n}$ coincides with $[N] \otimes F_f$, where $\bar{f} : W_f \to N$ is the continuous map that appears in the Stein factorization of f, $[N] \in H_n(N; \mathbb{Z}_2)$ is the fundamental class of N, and $F_f \in \mathfrak{N}_{m-n}$ is the unoriented cobordism class of a regular fiber of f. Note that when N is a closed manifold, the unoriented cobordism class F_f can be determined by the Stiefel-Whitney classes of M

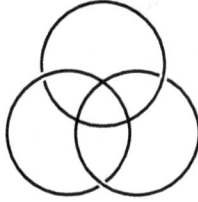

Fig. 4. The singular fiber that determines the cobordism class

together with $f^*(1^*) \in H^n(M; \mathbb{Z}_2)$, where $1^* \in H^n(N; \mathbb{Z}_2)$ is the Poincaré dual of the canonical generator $1 \in H_0(N; \mathbb{Z}_2)$. A similar remark is also valid in the oriented case. For details, see [12, §3].

Remark 3.2. Even if every component of every regular fiber is null-cobordant, the source manifold may not be null-cobordant.

For example, consider a C^∞ stable map $f: \mathbb{C}P^2 \to \mathbb{R}^3$. Every component of every regular fiber is diffeomorphic to S^1, which is null-cobordant. However, $\mathbb{C}P^2$ is not null-cobordant.

In fact, for a C^∞ stable map $f: M^4 \to \mathbb{R}^3$ of a closed oriented 4-dimensional manifold M^4, the cobordism class of M^4 is determined by singular fibers as depicted in Fig. 4 (see [13,14]).

Remark 3.3. Let M be a closed connected m-dimensional manifold. For a given Morse function $f': M \to \mathbb{R}$, we can modify it by homotopy so that we get an ordered Morse function $f: M \to \mathbb{R}$. Here, a Morse function f is *ordered* if for every pair of critical points p and q of f with index$(p) <$ index(q), we have $f(p) < f(q)$. As is observed in [4], if f is an ordered Morse function with $m \geq 3$, then every fiber of f is connected, and hence W_f is homeomorphic to a line segment. In this case, we have $\gamma_f = 0$, although the source manifold M may not be null-cobordant.

4. Further results

Let $f': M \to S^1$ be a continuous map of a smooth closed connected manifold M of dimension $m \geq 3$ into the circle. We assume that the induced homomorphism $f'_*: \pi_1(M) \to \pi_1(S^1)$ is surjective. Then, it is known that f' is homotopic to a Morse map $f: M \to S^1$ whose fibers are all non-empty and connected, where a *Morse map* is a smooth map whose critical points are all non-degenerate (for example, see [4, Theorem 1.3]). In this case, the quotient space W_f is canonically homeomorphic to S^1 through $\bar{f}$. In the following, a map is said to be *fiber-connected* if all of its fibers are non-empty and connected. For such a map, the quotient space is canonically homeomorphic to the target manifold.

Definition 4.1. Let $f_i \colon M_i \to N$ be smooth maps of closed m-dimensional manifolds M_i into a manifold N, $i = 0, 1$. We say that f_0 and f_1 are *bordant* if there exists a cobordism Q between M_0 and M_1 (i.e. Q is a compact $(m+1)$-dimensional manifold with ∂Q being identified with $M_0 \cup M_1$), and a smooth map $F \colon Q \to N \times [0, 1]$ such that $f_i = F|_{M_i} \colon M_i \to N \times \{i\}$, $i = 0, 1$. Such a map F is often called a *bordism* between f_0 and f_1. If everything is oriented, then we say that f_0 and f_1 are *oriented bordant* (for details, see [2], for example).

For example, if two smooth maps are homotopic, then they are bordant.

Proposition 4.2. *Let $f_i \colon M_i \to N$ be smooth maps of m-dimensional closed connected manifolds M_i into a connected n-dimensional manifold N with $m \geq n \geq 1$, $i = 0, 1$. We suppose that f_i are topologically stable and are fiber-connected, $i = 0, 1$. If f_0 and f_1 are bordant, then we have $\gamma_{f_0} = \gamma_{f_1} \in H_n(N; \mathfrak{N}_{m-n})$. If M_0, M_1 and N are oriented, and f_0 and f_1 are oriented bordant, then we have $\gamma_{f_0} = \gamma_{f_1} \in H_n(N; \Omega_{m-n})$.*

Proof. Let $F \colon Q \to N \times [0, 1]$ be the map as in Definition 4.1. By the Sard theorem, we can choose a point $y \in N$ which is a common regular value of f_0 and f_1. Then, by slightly perturbing F on the interior of Q, we may assume that F is transverse to the line segment $\{y\} \times [0, 1]$. Then, we see that $Q' = F^{-1}(\{y\} \times [0, 1])$ gives a cobordism between regular fibers of f_0 and f_1. Since, for each $i = 0, 1$, the regular fibers of f_i are all cobordant, we get the required result. $\square$

The above proposition implies that if two topologically stable fiber-connected maps $f_i \colon M_i \to N$, $i = 0, 1$, satisfy $\gamma_{f_0} \neq \gamma_{f_1}$, then they are not bordant. In particular, when $M_0 = M_1$, f_0 and f_1 cannot be homotopic.

Remark 4.3. Suppose that $f_i \colon M_i \to N$, $i = 0, 1$, are bordant as in Definition 4.1. We further assume that f_i are topologically stable, $i = 0, 1$, and that γ_{f_0} does not vanish. Under these assumptions, we have a sufficient condition for the non-vanishing of γ_{f_1} as follows.

Let $F \colon Q \to N \times [0, 1]$ be the smooth map that gives a bordism between f_0 and f_1. Let y be a regular value of f_0 such that $f_0^{-1}(y)$ contains a component which is not null-cobordant. Let $\alpha \colon [0, 1] \to N \times [0, 1]$ be a smooth embedding with $\alpha(0) = y \times \{0\}$ such that $\alpha(1) \in N \times \{1\}$ corresponds to a regular value of f_1 and that α is transverse to F. Let R be the component of $F^{-1}(\alpha([0,1]))$ which contains the component of $f_0^{-1}(y)$ which is not null-cobordant. By choosing α and F generic enough, we may further assume that the map $h = \alpha^{-1} \circ F \colon R \to [0,1]$ is a Morse function. If this Morse function has no critical points of index 1, then we can show that γ_{f_1} does not vanish.

Proposition 4.4. *Let m be an integer with $m \geq 3$. For an arbitrary $c \in \mathfrak{N}_m$ and for an arbitrary finite number of elements $c_j \in \mathfrak{N}_{m-1}$, $j = 1, 2, \ldots, k$, there exist a smooth closed connected m-dimensional manifold M and fiber-connected Morse maps $f_j \colon M \to S^1$ such that $[M] = c$ and $\gamma_{f_j} \in H_1(S^1; \mathfrak{N}_{m-1}) \cong \mathfrak{N}_{m-1}$ corresponds to c_j, $j = 1, 2, \ldots, k$.*

We also have a corresponding proposition for maps between oriented manifolds.

Proof of Proposition 4.4. Take a closed connected $(m-1)$-dimensional manifold F_j in the cobordism class c_j for each j, and a closed connected m-dimensional manifold M' with $[M'] = c$. Let us consider the closed connected m-dimensional manifold M given by

$$M = M' \natural \left(\natural_{j=1}^k (S^1 \times F_j) \right).$$

Since $S^1 \times F_j$ bounds $D^2 \times F_j$, it is null-cobordant, and hence we have $[M] = [M'] = c$.

Note that for each j, M naturally decomposes as $((S^1 \times F_j) \smallsetminus \mathrm{Int}\, D^m) \cup M'_j$, where $(S^1 \times F_j) \smallsetminus \mathrm{Int}\, D^m$ and M'_j are attached along their sphere boundaries. Let us construct a continuous map $f'_j \colon M \to S^1$ as follows. On $(S^1 \times F_j) \smallsetminus \mathrm{Int}\, D^m$, it is homotopic to the restriction of the projection $S^1 \times F_j \to S^1$ to the first factor and is a constant map on the boundary of $(S^1 \times F_j) \smallsetminus \mathrm{Int}\, D^m$. We define f'_j on M'_j to be the constant map to the same value in S^1. We may assume that f'_j is smooth on $(S^1 \times F_j) \smallsetminus \mathrm{Int}\, D^m$ and it has $\{*\} \times F_j$ as a regular fiber. Then by [4, Theorem 1.3], f'_j is homotopic to a fiber-connected Morse map $f_j \colon M \to S^1$. By construction, the regular fibers of f_j are all cobordant to F_j. Hence, f_j satisfy the desired properties. $\square$

Proposition 4.4 shows that at least for fiber-connected maps f, γ_f carries no information on the cobordism class of the source manifold.

We end this paper by posing some problems.

Problem 4.5. *(1) How about the case of maps of manifolds with non-empty boundaries?*

(2) By associating an "invariant" of a (regular or singular) fiber component corresponding to certain dimensional simplices of W_f, can we define a homology class of W_f?

(3) Study such kind of homology classes and their relations to the geometry and topology of the manifolds and the map. For example, can we define Vassiliev type invariants for maps using such homology classes?

Acknowledgements

The authors would like to thank the anonymous referee for stimulating questions and comments. The propositions in §4 were added after the referee's question.

The first author has been supported in part by CAPES, Brazil. The second author has been supported in part by JSPS KAKENHI Grant Number 23244008, 23654028.

References

1. O. Burlet and G. de Rham, *Sur certaines applications génériques d'une variété close à 3 dimensions dans le plan*, Enseignement Math. (2) **20** (1974), 275–292.
2. P.E. Conner and E.E. Floyd, *Differentiable periodic maps*, Ergebnisse der Mathematik und ihrer Grenzgebiete, N. F., Band 33, Academic Press Inc., Publishers, New York; Springer-Verlag, Berlin-Göttingen-Heidelberg, 1964.
3. S.K. Donaldson and P.B. Kronheimer, *The geometry of four-manifolds*, Oxford Mathematical Monographs, Oxford Science Publications, The Clarendon Press, Oxford University Press, New York, 1990.
4. D.T. Gay and R. Kirby, *Indefinite Morse 2-functions; broken fibrations and generalizations*, arXiv:1102.0750v2 [math.GT].
5. J.T. Hiratuka and O. Saeki, *Triangulating Stein factorizations of generic maps and Euler characteristic formulas*, RIMS Kôkyûroku Bessatsu, B38 (2013), 61–89.
6. J.F.P. Hudson, *Piecewise linear topology*, W.A. Benjamin, Inc., New York, 1969.
7. M. Kobayashi and O. Saeki, *Simplifying stable mappings into the plane from a global viewpoint*, Trans. Amer. Math. Soc. **348** (1996), 2607–2636.
8. L. Kushner, H. Levine and P. Porto, *Mapping three-manifolds into the plane I*, Boletin de la Sociedad Matemática Mexicana **29** (1984), 11–31.
9. H. Levine, *Classifying immersions into $\mathbb{R}^4$ over stable maps of 3-manifolds into $\mathbb{R}^2$*, Lecture Notes in Math., Vol. 1157, Springer-Verlag, Berlin, New York, 1985.

10. W. Motta, P. Porto Jr. and O. Saeki, *Stable maps of 3-manifolds into the plane and their quotient spaces*, Proc. London Math. Soc. (3) **71** (1995), 158–174.

11. P. Porto Jr and Y.K.S. Furuya, *On special generic maps from a closed manifold into the plane*, Topology Appl. **35** (1990), 41–52.

12. O. Saeki, *Studying the topology of Morin singularities from a global viewpoint*, Math. Proc. Camb. Phil. Soc. **117** (1995) 223–235.

13. O. Saeki, *Singular fibers and 4-dimensional cobordism group*, Pacific J. Math. **248** (2010), 233–256.

14. O. Saeki and T. Yamamoto, *Singular fibers of stable maps and signatures of 4-manifolds*, Geom. Topol. **10** (2006), 359–399.

15. K. Sakuma, *On the topology of simple fold maps*, Tokyo J. Math. **17** (1994), 21–31.

16. R. Thom, *Quelques propriétés globales des variétés différentiables*, Comment. Math. Helv. **28** (1954), 17–86.

17. C.T.C. Wall, *Determination of the cobordism ring*, Ann. of Math. **72** (1960), 292–311.

The reconstruction and recognition problems for homogeneous hypersurface singularities

A. V. Isaev

*Department of Mathematics, The Australian National University,
Canberra, ACT 0200, Australia*
`alexander.isaev@anu.edu.au`

By the well-known Mather-Yau theorem, a complex hypersurface germ V with isolated singularity is fully determined by its moduli algebra $\mathcal{A}(V)$. The proof of this theorem does not provide an explicit procedure for recovering V from $\mathcal{A}(V)$, and finding such a procedure is a long-standing open question, called the reconstruction problem. In the present paper we survey a recently proposed method for reconstructing V from $\mathcal{A}(V)$ up to biholomorphic equivalence under the assumption that the singularity of V is homogeneous, in which case $\mathcal{A}(V)$ coincides with the Milnor algebra of V. As part of our discussion of the method, we give a characterization of the algebras arising from finite polynomial maps with homogeneous components of equal degrees. For gradient maps, the question of describing such algebras is a special case of the so-called recognition problem for the moduli algebras of general isolated hypersurface singularities.

Keywords: isolated hypersurface singularities, the Mather-Yau theorem, Milnor algebras
AMS classification numbers: 32S25, 13H10

1. Introduction

Let $\mathcal{O}_n$ be the local complex algebra of holomorphic function germs at the origin in $\mathbb{C}^n$ with $n \geq 2$. For a hypersurface germ V at the origin (considered with its reduced complex structure) denote by $I(V)$ the ideal of elements of $\mathcal{O}_n$ that vanish on V. Fix a generator f of $I(V)$ and let $\mathcal{A}(V)$ be the quotient of $\mathcal{O}_n$ by the ideal generated by f and all its first-order partial derivatives. The algebra $\mathcal{A}(V)$, called the *moduli algebra* or *Tjurina algebra* of V, is in fact independent of the choice of f as well as the coordinate system near the origin. Furthermore, the moduli algebras of biholomorphically equivalent hypersurface germs are isomorphic. It is clear that $\mathcal{A}(V)$ is non-zero if and only if V is singular. We assume that $0 < \dim_{\mathbb{C}} \mathcal{A}(V) < \infty$, which occurs if and only if the singularity of V is isolated (see, e.g. Chapter 1 in [GLS]).

By the well-known Mather-Yau theorem (see [MY]), two hypersurface germs V_1, V_2 in $\mathbb{C}^n$ with isolated singularities are biholomorphically equivalent if their moduli algebras $\mathcal{A}(V_1)$, $\mathcal{A}(V_2)$ are isomorphic. Thus, given the dimension n, the moduli algebra $\mathcal{A}(V)$ determines V up to biholomorphism. For example, if $\dim_{\mathbb{C}} \mathcal{A}(V) = 1$, then V is biholomorphic to the germ of the hypersurface $\{z_1^2 + \cdots + z_n^2 = 0\}$, and if $\dim_{\mathbb{C}} \mathcal{A}(V) = 2$, then V is biholomorphic to the germ of the hypersurface $\{z_1^2 + \cdots + z_{n-1}^2 + z_n^3 = 0\}$. The proof of the Mather-Yau theorem does not provide an explicit procedure for recovering the germ V from the algebra $\mathcal{A}(V)$ in general, and finding a way for reconstructing V (or at least some invariants of V) from $\mathcal{A}(V)$ is an interesting open problem, called the *reconstruction problem* (cf. [Y1], [Y2], [S]). Recently, we have proposed a method for restoring V from $\mathcal{A}(V)$ up to biholomorphic equivalence under the assumption that the singularity of V is homogeneous (see [IK]). In the present paper we discuss this method and related questions.

Let V be a hypersurface germ having an isolated singularity. The singularity of V is said to be *homogeneous* if for some (hence for every) generator f of $I(V)$ there is a coordinate system near the origin in which f becomes the germ of a homogeneous polynomial. In this case f lies in the Jacobian ideal $\mathcal{J}(f)$ in $\mathcal{O}_n$, which is the ideal generated by all first-order partial derivatives of f. Hence, for a homogeneous singularity, $\mathcal{A}(V)$ coincides with the *Milnor algebra* $\mathcal{O}_n/\mathcal{J}(f)$ for any generator f of $I(V)$.

Now, we let $Q(z)$, with $z := (z_1, \ldots, z_n)$, be a holomorphic $(m+1)$-form on $\mathbb{C}^n$, i.e. a homogeneous polynomial of degree $m+1$ in the variables $z_1, \ldots, z_n$, where $m \geq 2$. Consider the germ V_Q of the hypersurface $\{Q(z) = 0\}$ and assume that: (i) the singularity of V_Q is isolated, and (ii) the germ of Q generates $I(V_Q)$. These two conditions are equivalent to the nonvanishing of the discriminant $\Delta(Q)$ of Q (see Chapter 13 in [GKZ]). The method proposed in [IK] recovers the form Q (hence the germ V_Q) from the algebra $\mathcal{A}(V_Q)$ up to linear equivalence, where two forms Q_1, Q_2 on $\mathbb{C}^n$ are called linearly equivalent if there exists a non-degenerate linear transformation L of $\mathbb{C}^n$ such that $Q_2 = Q_1 \circ L$.

We review the method of [IK] in Section 2. Define $\mathbf{Q}$ to be the gradient map $\mathbf{Q} : \mathbb{C}^n \to \mathbb{C}^n$, $z \mapsto \operatorname{grad} Q(z)$. The condition $\Delta(Q) \neq 0$ means that the fiber $\mathbf{Q}^{-1}(0)$ consists of 0 alone, i.e. $\mathbf{Q}$ is finite at the origin. The main content of this method is recovery of the map $\mathbf{Q}$ from $\mathcal{A}(V_Q)$ up to linear equivalence, where we say that two maps $\Phi_1, \Phi_2 : \mathbb{C}^n \to \mathbb{C}^n$ are linearly equivalent if there exist non-degenerate linear transformations L_1, L_2 of $\mathbb{C}^n$ such that $\Phi_2 = L_1 \circ \Phi_1 \circ L_2$. In fact, in [IK] we consider a more general situation.

Let $p_1, \ldots, p_n$, be holomorphic m-forms on $\mathbb{C}^n$ and I the ideal in $\mathcal{O}_n$ generated by the germs of these forms at the origin. Define $\mathbf{P}$ to be the map of $\mathbb{C}^n$ given by $z \mapsto (p_1(z), \ldots, p_n(z))$ and set $\mathcal{A}_{\mathbf{P}} := \mathcal{O}_n/I$. We assume that $\dim_{\mathbb{C}} \mathcal{A}_{\mathbf{P}} < \infty$, which occurs if and only if $\mathbf{P}$ is finite at the origin, i.e. $\mathbf{P}^{-1}(0) = \{0\}$ (see Chapter 1 in [GLS]). The latter condition is equivalent to the non-vanishing of the resultant of $p_1, \ldots, p_n$ (see Chapter 13 in [GKZ]). In [IK] we proposed a procedure (which requires only linear-algebraic manipulations) for explicitly recovering the map $\mathbf{P}$ from $\mathcal{A}_{\mathbf{P}}$ up to linear equivalence. Applying this procedure to the algebra $\mathcal{A}(\mathcal{V}_Q)$ arising from Q, one obtains a map $\mathbf{Q}'$ linearly equivalent to $\mathbf{Q}$. It is then not hard to derive from $\mathbf{Q}'$ an $(m+1)$-form Q' linearly equivalent to Q.

A priori, one can attempt to apply the procedure of [IK] to any complex commutative associative finite-dimensional algebra. Therefore, a natural problem is to characterize the algebras isomorphic to $\mathcal{A}_{\mathbf{P}}$ for some $\mathbf{P}$ as above. In the case of gradient maps, this characterization problem is a special case of the well-known *recognition problem* for the moduli algebras of general isolated hypersurface singularities and the corresponding Lie algebras of derivations (see, e.g. [Y1], [Y2], [S]). In Section 2 we give an elementary solution of the recognition problem for the algebras $\mathcal{A}_{\mathbf{P}}$ prior to our review of the reconstruction method of [IK].

We conclude the paper by giving an example of application of the method of [IK] to the Milnor algebras of simple elliptic singularities of type $\tilde{E}_7$ in Section 3. An analogous example for the Milnor algebras of simple elliptic singularities of type $\tilde{E}_6$ was discussed in [IK].

Acknowledgement. This work is supported by the Australian Research Council.

2. The reconstruction method

For fixed $m, n \geq 2$ let $\mathbf{P} : \mathbb{C}^n \to \mathbb{C}^n$ be a map that is finite at the origin and whose components $p_1, \ldots, p_n$ are holomorphic m-forms on $\mathbb{C}^n$. Define I to be the ideal in $\mathcal{O}_n$ generated by the germs of these forms at the origin and set $\mathcal{A}_{\mathbf{P}} := \mathcal{O}_n/I$. Since $m \geq 2$, one has $\dim_{\mathbb{C}} \mathcal{A}_{\mathbf{P}} \geq n + 1 > 1$. Before reviewing the method of [IK] for recovering the map $\mathbf{P}$ from $\mathcal{A}_{\mathbf{P}}$, we will give a simple characterization of the algebras that arise from finite homogeneous polynomial maps as above among all complex commutative associative finite-dimensional algebras.

Let $\mathcal{A}$ be a local unital complex commutative associative algebra with $1 < \dim_{\mathbb{C}} \mathcal{A} < \infty$. We say that $\mathcal{A}$ has Property $(*)$ if it satisfies conditions $(*)_1$–$(*)_3$ stated below.

$(*)_1$: $\mathcal{A}$ is a *standard graded algebra*, i.e. one has

$$A = \bigoplus_{i \geq 0} \mathcal{L}_i,$$

where $\mathcal{L}_i$ are linear subspaces of $\mathcal{A}$ with $\mathcal{L}_0 \simeq \mathbb{C}$, $\mathcal{L}_i \mathcal{L}_j \subset \mathcal{L}_{i+j}$ for all i, j, and $\mathcal{L}_l = \mathcal{L}_1^l$ for all $l \geq 1$. In this case, if $\mathfrak{m}$ is the (unique) maximal ideal of $\mathcal{A}$, then

$$\mathfrak{m} = \bigoplus_{i \geq 1} \mathcal{L}_i,$$

and $\mathcal{L}_i$ is a complement to $\mathfrak{m}^{i+1}$ in $\mathfrak{m}^i$ for all $i \geq 0$, where $\mathfrak{m}^0 := \mathcal{A}$. Hence $\mathcal{A}$ is a standard graded algebra if and only if $\mathcal{A}$ is isomorphic to its *associated graded algebra*

$$\mathrm{Gr}(\mathcal{A}) := \bigoplus_{i=0}^{\nu} \mathfrak{m}^i/\mathfrak{m}^{i+1},$$

where ν is the nil-index of $\mathfrak{m}$, i.e. the largest integer μ with $\mathfrak{m}^\mu \neq 0$ (note that $\mathfrak{m}$ is a nilpotent algebra by Nakayama's lemma).

$(*)_2$: $\mathcal{A}$ is a complete intersection.

$(*)_3$: for some $M \geq 2$ the following holds:

$$\dim_\mathbb{C} \mathfrak{m}^i/\mathfrak{m}^{i+1} = \dim_\mathbb{C} \mathcal{P}_i^N \quad \text{for } i = 1, \ldots, M-1,$$

$$\dim_\mathbb{C} \mathfrak{m}^M/\mathfrak{m}^{M+1} = \dim_\mathbb{C} \mathcal{P}_M^N - N,$$

where $N := \dim_\mathbb{C} \mathfrak{m}/\mathfrak{m}^2$ is the embedding dimension of $\mathcal{A}$ and for every $i \geq 1$ we denote by $\mathcal{P}_i^N$ the vector space of all i-forms on $\mathbb{C}^N$ (observe that $N > 0$). Note that the numbers $\dim_\mathbb{C} \mathcal{P}_i^N$ are well-known:

$$\dim_\mathbb{C} \mathcal{P}_i^N = \binom{i + N - 1}{i}.$$

Remark 2.1. Verification of conditions $(*)_2$ and $(*)_3$ for a given algebra $\mathcal{A}$ is not hard. Indeed, $(*)_2$ means that

$$\dim_\mathbb{C} H_1(K_{f_1, \ldots, f_N}) = N, \tag{2.1}$$

where $K_{f_1, \ldots, f_N}$ is the Koszul complex constructed from a basis $f_1, \ldots, f_N$ in a complement to $\mathfrak{m}^2$ in $\mathfrak{m}$ (see, e.g. pp. 169–172 in [M]); computing the left-hand side of (2.1) is straightforward from the definitions. Further, computation of all the dimensions $\dim_\mathbb{C} \mathfrak{m}^i/\mathfrak{m}^{i+1}$ required for verifying $(*)_3$

is easily done as well. In contrast, verification of condition $(*)_1$ may be quite difficult, especially if one is interested in constructing a standard grading on $\mathcal{A}$ explicitly. For example, to deal with condition $(*)_1$, one can first check whether $\mathrm{Gr}(\mathcal{A})$ is Gorenstein and, if this is the case, use the criterion for isomorphism of Gorenstein algebras obtained in [FIKK]. We note that a necessary condition for the existence of a standard grading is given in Proposition 8.1 in [FK].

We are now ready to state our result.

Theorem 2.2. Let $\mathcal{A}$ be a local unital complex commutative associative algebra with $1 < \dim_{\mathbb{C}} \mathcal{A} < \infty$. Then $\mathcal{A}$ is isomorphic to the algebra $\mathcal{A}_{\mathbf{P}}$ for some $\mathbf{P}$ if and only if $\mathcal{A}$ has Property $(*)$, in which case $N = n$ and $M = m = \nu/n + 1$.

Proof. First, we let $\mathcal{A} = \mathcal{A}_{\mathbf{P}}$ for some map $\mathbf{P}$ as above and show that $\mathcal{A}$ has Property $(*)$. Indeed, $p_1, \ldots, p_n$ is a regular sequence in $\mathcal{O}_n$ (see Theorem 2.1.2 in [BH]), hence $\mathcal{A}$ satisfies $(*)_2$. Next, we clearly have $N = n$, and for any $i \geq 0$ define $\mathcal{L}_i$ to be the linear subspace of $\mathcal{A}$ that consists of all elements represented by germs of forms in $\mathcal{P}_i^n$. These subspaces form a standard grading on $\mathcal{A}$, hence $(*)_1$ is satisfied. Further, condition $(*)_3$ obviously holds with $M = m$. Finally, as we noted in [IK], the embedding dimension n divides ν and one has $m = \nu/n + 1$.

Conversely, let $\mathcal{A}$ be a local unital complex commutative associative algebra with $1 < \dim_{\mathbb{C}} \mathcal{A} < \infty$ having Property $(*)$. Choose a basis $f_1, \ldots, f_N$ in a complement to $\mathfrak{m}^2$ in $\mathfrak{m}$. The elements $f_1, \ldots, f_N$ generate $\mathcal{A}$ (as an algebra), hence $\mathcal{A}$ is isomorphic to $\mathbb{C}[z_1, \ldots, z_N]/R$, where R is the ideal of all relations, i.e. polynomials $P \in \mathbb{C}[z_1, \ldots, z_N]$ with $P(f_1, \ldots, f_N) = 0$. Observe that R contains all homogeneous polynomials of degree greater than ν. It then follows that $\mathcal{A}$ is isomorphic to $\mathcal{O}_N/R_0$, where R_0 is the ideal in $\mathcal{O}_N$ generated by the germs of all relations.

We will now use condition $(*)_1$. Namely, choosing $f_1, \ldots, f_N$ to be a basis in $\mathcal{L}_1$, we see that R is generated by finitely many homogeneous relations, which yields $R_0 \subset \mathfrak{m}_N^2$, where $\mathfrak{m}_N$ is the maximal ideal of $\mathcal{O}_N$. Next, since $\mathcal{A}$ is isomorphic to the quotient of the regular local ring $\mathcal{O}_N$ by the ideal $R_0 \subset \mathfrak{m}_N^2$, condition $(*)_2$ means that the minimal number of generators of R_0 is equal to N (see, e.g. pp. 170–171 in [M]). Further, condition $(*)_3$ means that the ideal R contains no relations of degree less than M and that the linear subspace $\mathcal{H} \subset R$ of all homogeneous relations of degree M has dimension N.

Choose a set of generators of R_0 consisting of N elements, let $g_1, \ldots, g_N$ be holomorphic functions representing the generators, and fix a basis $p_1, \ldots, p_N$ of $\mathcal{H}$. Then in a neighborhood of the origin we have

$$p_i = \sum_{j=1}^{N} a_{ij} g_j, \quad i = 1, \ldots, N \tag{2.2}$$

for some holomorphic functions a_{ij}. On the other hand, if we complete the basis $p_1, \ldots, p_N$ to a set of generators of R by adding homogeneous elements $q_1, \ldots, q_L \in R$ of degree greater than M, each function g_j can be written as

$$g_j = \sum_{k=1}^{N} b_{jk} p_k + \sum_{\ell=1}^{L} c_{j\ell} q_\ell, \quad j = 1, \ldots, N, \tag{2.3}$$

where b_{jk}, $c_{j\ell}$ are holomorphic functions. Plugging (2.3) into (2.2) and collecting terms of degree M in both sides of the resulting identity, we see that the matrix $(a_{ij}(0))$ is non-degenerate. Letting (d_{jk}) be the inverse of (a_{ij}) near the origin (with $d_{jk}(0) = b_{jk}(0)$), from equation (2.2) we then obtain

$$g_j = \sum_{k=1}^{N} d_{jk} p_k, \quad j = 1, \ldots, N,$$

which implies that the germs of $p_1, \ldots, p_N$ generate R_0.

Thus, we have shown that the germs of the elements of every basis of $\mathcal{H}$ generate R_0. Let $p_1, \ldots, p_N$ be any basis and $\mathbf{P} := (p_1, \ldots, p_N)$. By the condition $\dim_{\mathbb{C}} \mathcal{A} < \infty$, the map $\mathbf{P}$ is finite at the origin. We have thus proved that $\mathcal{A}$ is isomorphic to $\mathcal{A}_{\mathbf{P}}$. $\qquad\square$

The proof of Theorem 2.2 provides a method for producing a map $\mathbf{P}$ from any algebra $\mathcal{A}$ having Property $(*)$, but this method requires that a standard grading on $\mathcal{A}$ be explicitly defined. As we noted in Remark 2.1, finding such a grading may be hard. On the other hand, in article [IK] we proposed an algorithm for recovering $\mathbf{P}$ up to linear equivalence from the algebra $\mathcal{A}_{\mathbf{P}}$, where one does not need to know a standard grading explicitly. Assuming that $\mathcal{A}_{\mathbf{P}}$ is given as an abstract algebra (i.e. by a multiplication table with respect to some basis), we summarize the main steps of the algorithm as follows (see [IK] for details):

1. Find $\mathfrak{m}$ and its nil-index ν.

2. Determine n from the formula $n = \dim_{\mathbb{C}} \mathfrak{m}/\mathfrak{m}^2$.

3. Determine m from the formula $m = \nu/n + 1$.

4. Choose a complement to $\mathfrak{m}^2$ in $\mathfrak{m}$ and an arbitrary basis $f_1, \ldots, f_n$ in this complement.

5. Calculate $q_1(f), \ldots, q_K(f)$, where $f := (f_1, \ldots, f_n)$, $q_1(z), \ldots, q_K(z)$ are all monomials of degree m in $z := (z_1, \ldots, z_n)$, and $K := \dim_{\mathbb{C}} \mathcal{P}_m^n$.

6. Choose a complement $\mathcal{S}$ to $\mathfrak{m}^{m+1}$ in $\mathfrak{m}^m$.

7. Compute $\pi(q_1(f)), \ldots, \pi(q_K(f))$, where $\pi : \mathfrak{m}^m \to \mathcal{S}$ is the projection onto $\mathcal{S}$ with kernel $\mathfrak{m}^{m+1}$.

8. Find n linearly independent linear relations among the vectors $\pi(q_1(f)), \ldots, \pi(q_K(f))$:

$$\sum_{\rho=1}^{K} \gamma_{\sigma\rho} \pi(q_\rho(f)) = 0, \quad \sigma = 1, \ldots, n, \quad \gamma_{\sigma\rho} \in \mathbb{C}.$$

9. The following formula then gives a map linearly equivalent to $\mathbf{P}$:

$$\Phi : \mathbb{C}^n \to \mathbb{C}^n, \quad z \mapsto \Gamma q(z),$$

where $\Gamma := (\gamma_{\sigma\rho})_{\sigma=1,\ldots,n, \; \rho=1,\ldots,K}$, and $q := (q_1, \ldots, q_K)$.

Remark 2.3. The algorithm of [IK] shows, in particular, that if two algebras $\mathcal{A}_{\mathbf{P}}$ and $\mathcal{A}_{\mathbf{P}'}$ are isomorphic, then the maps $\mathbf{P}$ and $\mathbf{P}'$ are linearly equivalent. Hence, the sufficiency implication of Theorem 2.2 can be strengthened as follows: for every algebra $\mathcal{A}$ having Property $(*)$ there exist exactly one, up to linear equivalence, map $\mathbf{P}$ such that $\mathcal{A}$ is isomorphic to $\mathcal{A}_{\mathbf{P}}$.

One can apply the above algorithm to $\mathbf{P} = \mathbf{Q} := \operatorname{grad} Q$ for a holomorphic $(m+1)$-form Q on $\mathbb{C}^n$ with $\Delta(Q) \neq 0$. Let Φ be a map linearly equivalent to $\mathbf{Q}$ derived from the algebra $\mathcal{A}(\mathcal{V}_Q)$, where $\mathcal{V}_Q$ is the germ of the hypersurface $\{Q = 0\}$ at the origin. For any $n \times n$-matrix D we now introduce the holomorphic differential 1-form $\omega^{\Phi^D} := \sum_{i=1}^{n} \Phi_i^D dz_i$ on $\mathbb{C}^n$, where $\Phi_1^D, \ldots, \Phi_n^D$ are the components of the map $\Phi^D := D\Phi$. Consider the equation

$$d\omega^{\Phi^D} = 0 \tag{2.4}$$

as a linear system with respect to the entries of the matrix D. It is explained in [IK] that in order to recover Q from $\mathcal{A}(\mathcal{V}_Q)$ up to linear equivalence one needs to complement the above algorithm with the following two steps:

10. Find a matrix $D \in \mathrm{GL}(n, \mathbb{C})$ satisfying system (2.4).

11. Integrate Φ^D to obtain an $(m+1)$-form linearly equivalent to Q.

3. An example of application of the reconstruction method

We will now illustrate our reconstruction method, as well as Property $(*)$, by an example of a 1-parameter family of algebras. Namely, for $t \in \mathbb{C}$, $t \neq \pm 2$, let $\mathcal{A}_t$ be the complex commutative 9-dimensional algebra given with respect to a certain basis $e_1, \ldots, e_9$ by the following multiplication table:

$$e_1 e_j = e_j \ for j = 1, \ldots, 9, \ e_2^2 = e_6 - \frac{2}{3}e_7 + e_8 - \frac{1}{3}e_9,$$

$$e_2 e_3 = \frac{1}{3}e_7 + e_8 - \frac{1}{3}e_9, \ e_2 e_4 = e_5 - \frac{1}{3}e_7 + \frac{1}{3}e_9, \ e_2 e_5 = \frac{2}{3}e_7 + 3e_8 + \frac{1}{3}e_9,$$

$$e_2 e_6 = -\frac{t}{6}e_7 + \frac{t}{6}e_9, \ e_2 e_7 = e_8, \ e_2 e_8 = 0, \ e_2 e_9 = -2e_8, \ e_3^2 = -\frac{t}{2}e_8,$$

$$e_3 e_4 = -\frac{t}{3}e_7 - \frac{t}{6}e_9, \ e_3 e_j = 0, \ j = 5,7,8,9, \ e_3 e_6 = e_8, \tag{3.1}$$

$$e_4^2 = e_3 - \frac{1}{3}e_7 + 2e_8 + \frac{1}{3}e_9, \ e_4 e_5 = \frac{1}{3}e_7 - \frac{1}{3}e_9, \ e_4 e_6 = \frac{2}{3}e_7 + 3e_8 + \frac{1}{3}e_9,$$

$$e_4 e_7 = e_8, \ e_4 e_8 = 0, \ e_4 e_9 = e_8, \ e_5^2 = e_8, \ e_5 e_j = 0, \ j = 6,7,8,9, \ e_6^2 = -\frac{t}{2}e_8,$$

$$e_6 e_j = 0, \ j = 7,8,9, \ e_7 e_j = 0, \ j = 7,8,9, \ e_8 e_j = 0, \ j = 8,9, \ e_9^2 = 0.$$

As we will see later, every algebra $\mathcal{A}_t$ has Property $(*)$. It is clear from (3.1) that $e_1 = 1$ (the identity element) and $\mathcal{A}_t$ is a local algebra with maximal ideal $\mathfrak{m}_t = \langle e_2, \ldots, e_9 \rangle$, where $\langle \cdot \rangle$ denotes linear span. We then have $\mathfrak{m}_t^2 = \langle e_3, e_5, e_6, e_7, e_8, e_9 \rangle$, $\mathfrak{m}_t^3 = \langle e_7, e_8, e_9 \rangle$, $\mathfrak{m}_t^4 = \langle e_8 \rangle$, $\mathfrak{m}_t^5 = 0$, hence $\nu = 4$. Further, by the formula at Step 2 of the algorithm given in Section 2, we obtain $n = 2$, which together with formula at Step 3 yields $m = 3$. We now list all monomials of degree 3 in $z := (z_1, z_2)$ as follows:

$$q_1(z) := z_1^3, \quad q_2(z) := z_1^2 z_2, \quad q_3(z) := z_1 z_2^2, \quad q_4(z) := z_2^3$$

(here $K = 4$). Next, we let $f_1 := e_2$, $f_2 := e_4$, which for $f := (f_1, f_2)$ yields

$$q_1(f) = -\frac{t}{6}e_7 + \frac{t}{6}e_9, \; q_2(f) = \frac{2}{3}e_7 + 2e_8 + \frac{1}{3}e_9,$$

$$q_3(f) = \frac{1}{3}e_7 - \frac{1}{3}e_9, \quad q_4(f) = -\frac{t}{3}e_7 - \frac{t}{6}e_9.$$

Further, define $\mathcal{S} := \langle e_7, e_9 \rangle$. Clearly, $\mathcal{S}$ is a complement to $\mathfrak{m}_t^4$ in $\mathfrak{m}_t^3$. Then for the projection $\pi : \mathfrak{m}_t^3 \to \mathcal{S}$ with kernel $\mathfrak{m}_t^4$ one has

$$\pi(q_1(f)) = -\frac{t}{6}e_7 + \frac{t}{6}e_9, \; \pi(q_2(f)) = \frac{2}{3}e_7 + \frac{1}{3}e_9,$$

$$\pi(q_3 f)) = \frac{1}{3}e_7 - \frac{1}{3}e_9, \quad \pi(q_4(f)) = -\frac{t}{3}e_7 - \frac{t}{6}e_9.$$

The vectors $\pi(q_1(f)), \pi(q_2(f)), \pi(q_3(f)), \pi(q_4(f))$ satisfy the following two linearly independent linear relations:

$$2\pi(q_1(f)) + t\pi(q_3(f)) = 0,$$

$$t\pi(q_2(f)) + 2\pi(q_4(f)) = 0.$$

Hence we have

$$\Gamma = \begin{pmatrix} 2 & 0 & t & 0 \\ 0 & t & 0 & 2 \end{pmatrix},$$

which for $q(z) := (q_1(z), q_2(z), q_3(z), q_4(z))$ yields

$$\Phi(z) = \Gamma q(z) = \begin{pmatrix} 2z_1^3 + tz_1 z_2^2 \\ tz_1^2 z_2 + 2z_2^3 \end{pmatrix}.$$

Further, for

$$D = \begin{pmatrix} d_{11} & d_{12} \\ d_{21} & d_{22} \end{pmatrix}$$

system (2.4) is equivalent to the following system of equations:

$$t(d_{11} - d_{22}) = 0,$$

$$td_{12} - 6d_{21} = 0, \tag{3.2}$$

$$6d_{12} - td_{21} = 0.$$

If $t \neq 0, \pm 6$, the only non-degenerate solutions of (3.2) are non-zero scalar matrices. Integrating Φ^D for such a matrix D we obtain a form proportional to

$$Q_t := z_1^4 + tz_1^2 z_2^2 + z_2^4.$$

If $t = 0$, any non-degenerate solution of (3.2) is a diagonal matrix with non-zero d_{11}, d_{22}. Integrating Φ^D for such a matrix D we obtain the form

$$\frac{1}{2}\left(d_{11}z_1^4 + d_{22}z_2^4\right),$$

which is linearly equivalent to

$$Q_0 := z_1^4 + z_2^4$$

by suitable dilations of the variables.

The remaining case $t = \pm 6$ is more interesting. In this situation D is a solution of (3.2) if and only if

$$d_{11} = d_{22}, \qquad d_{12} = \pm d_{21}.$$

Such a matrix D is non-degenerate if and only if $d_{11}^2 \mp d_{12}^2 \neq 0$. One obvious solution is given by $d_{11} = 1$, $d_{12} = 0$, in which case integrating Φ^D one obtains a form proportional to

$$Q_{\pm 6} := z_1^4 \pm 6z_1^2 z_2^2 + z_2^4$$

(observe that Q_6 and Q_{-6} are in fact linearly equivalent). If Q is a form arising from any other solution, then the algebras $\mathcal{A}(\mathcal{V}_Q)$ and $\mathcal{A}(\mathcal{V}_{Q_{\pm 6}})$ are isomorphic, which by the Mather-Yau theorem implies that $\mathcal{V}_Q$ and $\mathcal{V}_{Q_{\pm 6}}$ are biholomorphically equivalent hence Q and $Q_{\pm 6}$ are linearly equivalent. For example, letting $d_{11} = 0$, $d_{12} = 1$ one obtains

$$\Phi^D = \begin{pmatrix} \pm 6z_1^2 z_2 + 2z_2^3 \\ \pm 2z_1^3 + 6z_1 z_2^2 \end{pmatrix},$$

and integration of Φ^D leads to the form

$$\mathcal{Q}_{\pm 6} := \pm 2z_1^3 z_2 + 2z_1 z_2^3.$$

As explained above, by the Mather-Yau theorem each of $\mathcal{Q}_{\pm 6}$ is linearly equivalent to each of $Q_{\pm 6}$.

This last fact can also be understood without referring to the Mather-Yau theorem as follows. It is well-known that all non-equivalent binary quartics with non-vanishing discriminant are distinguished by the invariant

$$J := \frac{I_3^2}{\Delta},$$

where for any quartic

$$Q = a_4 z_1^4 + 4a_3 z_1^3 z_2 + 6a_2 z_1^2 z_2^2 + 4a_1 z_1 z_2^3 + a_0 z_2^4$$

one has

$$I_3(Q) = \det \begin{pmatrix} a_4 & a_3 & a_2 \\ a_3 & a_2 & a_1 \\ a_2 & a_1 & a_0 \end{pmatrix}$$

(see, e.g. pp. 28–29 in [O]). We then have $I(\mathcal{Q}_{\pm 6}) = I(\mathcal{Q}_{\pm 6}) = 0$, and therefore each of $\mathcal{Q}_{\pm 6}$ is linearly equivalent to each of $Q_{\pm 6}$ as stated above.

Thus, assuming that the algebra $\mathcal{A}_t$ has Property $(*)$, we have shown that it is isomorphic to $\mathcal{A}(\mathcal{V}_{Q_t})$ for all $t \neq \pm 2$. Note that $\mathcal{A}(\mathcal{V}_{Q_t})$ is a well-known family of algebras. Indeed, $\mathcal{A}(\mathcal{V}_{Q_t})$ is isomorphic to the moduli algebra $\mathcal{A}(\mathcal{V}_t)$ of the germ $\mathcal{V}_t$ of the following hypersurface in $\mathbb{C}^3$:

$$\left\{ (z_1, z_2, z_3) \in \mathbb{C}^3 : z_1^4 + t z_1^2 z_2^2 + z_2^4 + z_3^2 = 0 \right\}, \quad t \neq \pm 2.$$

These hypersurface singularities are called *simple elliptic singularities of type* $\tilde{E}_7$. We stress that, unlike $\mathcal{V}_{Q_t}$, these singularities are not homogeneous.

Observe that one basis in which the algebra $\mathcal{A}(\mathcal{V}_t)$ is given by multiplication table (3.1) is as follows:

$$e_1 = 1, \; e_2 = Z_1 + Z_1 Z_2, \; e_3 = Z_2^2 + Z_1 Z_2^2, \; e_4 = Z_2 + Z_1^2 Z_2,$$

$$e_5 = Z_1 Z_2 + 2 Z_1 Z_2^2, \; e_6 = Z_1^2 + 3 Z_1^2 Z_2, \; e_7 = Z_1^2 Z_2 + Z_1 Z_2^2,$$

$$e_8 = Z_1^2 Z_2^2, \; e_9 = Z_1^2 Z_2 - 2 Z_1 Z_2^2,$$

where Z_j is the element represented by the germ of the coordinate function z_j, $j = 1, 2$.

We will now independently check that $\mathcal{A}_t$ has Property $(*)$ for every $t \neq \pm 2$. First of all, the subspaces

$$\mathcal{L}_0 := \langle e_1 \rangle, \quad \mathcal{L}_1 := \left\langle e_2 - e_5 + \frac{2}{3} e_7 - \frac{2}{3} e_9, e_4 - \frac{2}{3} e_7 - \frac{1}{3} e_9 \right\rangle,$$

$$\mathcal{L}_2 := \left\langle e_3 - \frac{1}{3} e_7 + \frac{1}{3} e_9, e_5 - \frac{2}{3} e_7 + \frac{2}{3} e_9, e_6 - 2 e_7 - e_9 \right\rangle,$$

$$\mathcal{L}_3 := \langle e_7, e_9 \rangle, \quad \mathcal{L}_4 := \langle e_8 \rangle$$

form a standard grading on $\mathcal{A}_t$, thus $\mathcal{A}_t$ satisfies condition $(*)_1$. Further, we clearly have $N = 2$ and $\dim_{\mathbb{C}} \mathcal{P}_i^2 = i + 1$, which immediately yields that condition $(*)_3$ is satisfied with $M = 3$.

It remains to verify condition $(*)_2$. This condition means that

$$\dim_{\mathbb{C}} H_1(K_{f_1, f_2}) = 2, \tag{3.3}$$

86

where K_{f_1,f_2} is the Koszul complex constructed from any basis f_1, f_2 in a complement to $\mathfrak{m}_t^2$ in $\mathfrak{m}_t$. As before, we choose $f_1 = e_2$, $f_2 = e_4$, in which case (3.3) is equivalent to

$$\dim_{\mathbb{C}} \frac{\{(u,v) \in \mathcal{A}_t \times \mathcal{A}_t : ue_2 + ve_4 = 0\}}{\{(u,v) \in \mathcal{A}_t \times \mathcal{A}_t : u = -we_4, \, v = we_2 \text{ for some } w \in \mathcal{A}_t\}} = 2.$$

The above identity easily follows by expanding u, v, w with respect to the basis $e_1, \ldots, e_9$ and utilizing multiplication table (3.1). Thus, the algebra $\mathcal{A}_t$ has Property $(*)$ as stated.

References

BH.   Bruns, W. and Herzog, J., *Cohen-Macaulay Rings*, Cambridge Studies in Advanced Mathematics 39, Cambridge University Press, Cambridge, 1993.

FIKK.   Fels, G., Isaev, A., Kaup, W. and Kruzhilin, N., Isolated hypersurface singularities and special polynomial realizations of affine quadrics, *J. Geom. Analysis* 21 (2011), 767–782.

FK.   Fels, G. and Kaup, W., Nilpotent algebras and affinely homogeneous surfaces, *Math. Ann.* 353 (2012), 1315–1350.

GKZ.   Gelfand, I. M., Kapranov, M. M. and Zelevinsky, A. V., *Discriminants, Resultants and Multidimensional Determinants*, Modern Birkhäuser Classics, Birkhäuser Boston, Inc., Boston, MA, 2008.

GLS.   Greuel, G.-M., Lossen, C. and Shustin, E., *Introduction to Singularities and Deformations*, Springer Monographs in Mathematics, Springer, Berlin, 2007.

IK.   Isaev, A. V. and Kruzhilin, N. G., Explicit reconstruction of homogeneous isolated hypersurface singularities from their Milnor algebras, to appear in *Proc. Amer. Math. Soc.*

MY.   Mather, J. and Yau, S. S.-T., Classification of isolated hypersurface singularities by their moduli algebras, *Invent. Math.* 69 (1982), 243–251.

M.   Matsumura, H., *Commutative Ring Theory*, Cambridge Studies in Advanced Mathematics 8, Cambridge University Press, Cambridge, 1986.

O.   Olver, P., *Classical Invariant Theory*, London Mathematical Society Student Texts 44, Cambridge University Press, Cambridge, 1999.

S.   Schulze, M., A solvability criterion for the Lie algebra of derivations of a fat point, *J. Algebra* 323 (2010), 2916–2921.

Y1.   Yau, S. S.-T., Solvable Lie algebras and generalized Cartan matrices arising from isolated singularities, *Math. Z.* 191 (1986), 489–506.

Y2.   Yau, S. S.-T., Solvability of Lie algebras arising from isolated singularities and nonisolatedness of singularities defined by $\mathfrak{sl}(2,\mathbb{C})$ invariant polynomials, *Amer. J. Math.* 113 (1991), 773–778.

Openings of differentiable map-germs and unfoldings

Dedicated to Professor Satoshi Koike for his 60th birthday

Goo Ishikawa

Department of Mathematics, Hokkaido University, Japan
ishikawa@math.sci.hokudai.ac.jp

The algebraic notion of *openings* of a map-germ is introduced in this paper. An opening separates the self-intersections of the original map-germ, preserving its singularities. The notion of openings is different from the notion of unfoldings. Openings do not unfold the singularities. For example, the swallowtail is an opening of the Whitney cusp map-germ from plane to plane and the open swallowtail is a versal opening of them. Openings of map-germs appear as typical singularities in several problems of geometry and its applications. The notion of openings has close relations to isotropic map-germs in a symplectic space and integral map-germs in a contact space. We describe the openings of Morin singularities, namely, stable unfoldings of map-germs of corank one. The relation of unfoldings and openings are discussed. Moreover we provide a method to construct versal openings of map-germs and give versal openings of stable map-germs $(\mathbf{R}^4, 0) \to (\mathbf{R}^4, 0)$. Lastly the relation of lowerable vector fields and openings is discussed.

Keywords: versal opening, ramification module, lowerable vector field
AMS classification numbers: 58C27, 14E40, 32S45

1. Introduction

There is a sequence of well-known singularities of map-germs: The *Whitney cusp* $f : (\mathbf{R}^2, 0) \to (\mathbf{R}^2, 0)$, $f(x, u) = (x^3 + ux, u)$, the *swallowtail* $F : (\mathbf{R}^2, 0) \to (\mathbf{R}^3, 0)$, $F(x, u) = (f(x, u), x^4 + \frac{2}{3}ux^2)$, and the *open swallowtail* $\widetilde{F} : (\mathbf{R}^2, 0) \to (\mathbf{R}^4, 0)$, $\widetilde{F}(x, u) = (F(x, u), x^5 + \frac{4}{9}ux^3)$.

They have the same singular locus and the same kernel field of the differential along the singular locus, while the self-intersections are resolved.

What is the algebraic structure behind them? One of answers to the above question is presented in this paper. In fact we observe, for the swallowtail $F(x, u) = (x^3 + ux, u, x^4 + \frac{2}{3}ux^2)$, we see that

$$d(x^4 + \frac{2}{3}ux^2) = \frac{4}{3}x \, d(x^3 + ux) - \frac{4}{9}x^2 \, du \in \langle d(x^3 + ux), \, du \rangle_{\mathcal{E}_2}.$$

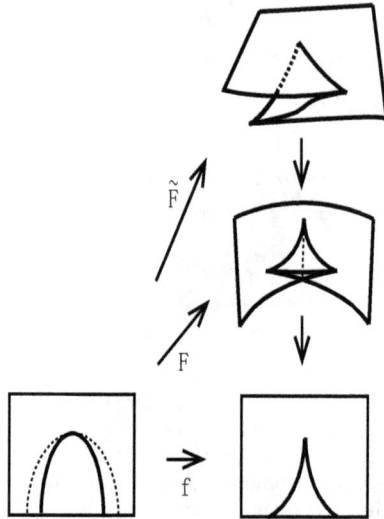

For the open swallowtail $\widetilde{F}(x, u) = (x^3 + ux, u, x^4 + \frac{2}{3}ux^2, x^5 + \frac{4}{9}ux^3)$, we have that

$$d(x^5 + \frac{4}{9}ux^3) = \frac{5}{3}x^2 \, d(x^3 + ux) - \frac{10}{9}x^3 \, du \in \langle d(x^3 + ux), \, du \rangle_{\mathcal{E}_2}.$$

Here d means the exterior differential and $\mathcal{E}_2$ denotes the **R**-algebra of C^∞ function-germs on $(\mathbf{R}^2, 0)$.

As the key construction, we introduce the notion of *openings* of multi-germs of mappings. To do this, first we summarise the auxiliary notions in this paper.

Let $f : (\mathbf{R}^n, A) \to (\mathbf{R}^m, b)$ be a multi-germ of a C^∞ map with $n \le m$. Here A is a finite subset of $\mathbf{R}^n$, $b \in \mathbf{R}^m$ and $f(A) = \{b\}$. We define the *Jacobi module* $\mathcal{J}_f$ of f by

$$\mathcal{J}_f = \{ \sum_{j=1}^{m} p_j \, df_j \mid p_j \in \mathcal{E}_{\mathbf{R}^n, A} \, (1 \le j \le m) \} \subset \Omega^1_{\mathbf{R}^n, A}$$

in the space $\Omega^1_{\mathbf{R}^n,A}$ of 1-form-germs on $(\mathbf{R}^n, A)$. Note that $\mathcal{J}_f$ is just the first order component of the graded differential ideal $\mathcal{J}_f^\bullet$ in $\Omega^\bullet_{\mathbf{R}^n,A}$ generated by $df_1, \ldots, df_m$. Then the singular locus, the non-immersive locus, of f is given by

$$\Sigma_f = \{x \in (\mathbf{R}^n, A) \mid \operatorname{rank} \mathcal{J}_f(x) < n\}.$$

Also we consider the *kernel field* $\operatorname{Ker}(f_* : T\mathbf{R}^n \to T\mathbf{R}^m)$ of the differential of f, along Σ_f.

For another map-germ $f' : (\mathbf{R}^n, A) \to (\mathbf{R}^{m'}, b')$, $n \leq m'$, if $\mathcal{J}_{f'} = \mathcal{J}_f$, then $\Sigma_{f'} = \Sigma_f$ and $\operatorname{Ker}(f'_*) = \operatorname{Ker}(f_*)$.

Then define the *ramification module* $\mathcal{R}_f$ of f by

$$\mathcal{R}_f = \{h \in \mathcal{E}_{\mathbf{R}^n,A} \mid dh \in \mathcal{J}_f\},$$

(cf. [12] [15]).

For $f : (\mathbf{R}^n, A) \to (\mathbf{R}^m, b), f' : (\mathbf{R}^n, A) \to (\mathbf{R}^{m'}, b')$, easily we see that $\mathcal{J}_{f'} = \mathcal{J}_f$ if and only if $\mathcal{R}_{f'} = \mathcal{R}_f$ (Lemma 2.1).

Definition 1.1. Given $h_1, \ldots, h_r \in \mathcal{R}_f$, the map-germ $F : (\mathbf{R}^n, A) \to \mathbf{R}^m \times \mathbf{R}^r = \mathbf{R}^{m+r}$ defined by

$$F = (f_1, \ldots, f_m, h_1, \ldots, h_r)$$

is called an *opening* of f, while f is called a *closing* of F.

Then, for any opening F of f, we have $\mathcal{R}_F = \mathcal{R}_f$, $\mathcal{J}_F = \mathcal{J}_f$, $\Sigma_F = \Sigma_f$ and $\operatorname{Ker}(F_*) = \operatorname{Ker}(f_*)$.

For example, the swallowtail is an opening of the Whitney cusp. The open swallowtail is an opening of the swallowtail and of the Whitney cusp. Note that an opening of an opening of f is an opening of f.

Definition 1.2. An opening $F = (f, h_1, \ldots, h_r)$ of f is called a *versal opening* (resp. a *mini-versal opening*) of $f : (\mathbf{R}^n, A) \to (\mathbf{R}^m, b)$, if $1, h_1, \ldots, h_r$ form a (minimal) system of generators of $\mathcal{R}_f$ as an $\mathcal{E}_{\mathbf{R}^m,b}$-module via $f^* : \mathcal{E}_{\mathbf{R}^m,b} \to \mathcal{E}_{\mathbf{R}^n,A}$.

Note that a versal opening of an opening of f is a versal opening of f. An opening of a versal opening of f is a versal opening of f.

A mini-versal opening $F : (\mathbf{R}^n, A) \to \mathbf{R}^{m+r}$ of f is unique up to left-equivalence and a versal opening $G : (\mathbf{R}^n, A) \to \mathbf{R}^{m+s}$ of f is left-equivalent to a mini-versal opening composed with an immersion $(\mathbf{R}^n, A) \to \mathbf{R}^{m+r} \hookrightarrow \mathbf{R}^{m+s}$ (Proposition 2.14).

Openings of map-germs appear as typical singularities in several problems of geometry and its applications. The openings naturally appear in the classification problem of "tangential singularities"([5] [16] [17]). Open swallowtails, open folded umbrellas, etc. appear as tangent varieties ([1]). We have applied opening constructions to the solution of the "stable"classification problem of tangent varieties to generic submanifolds in [17]. Moreover openings are related to singularities of isotropic mappings in symplectic spaces.

Let $T^*\mathbf{R}^n = \mathbf{R}^{2m}$ be the $2m$-dimensional symplectic space with the symplectic form $\omega = \sum_{i=1}^m dp_i \wedge dx_i$, and $f : (\mathbf{R}^n, A) \to \mathbf{R}^{2m}$ a multi-germ of isotropic mapping. Since $f^*\omega = 0$, we have that $\sum_{i=1}^m (p_i \circ f)d(x_i \circ f)$ is closed, so it is exact and there exists $e \in \mathcal{E}_{\mathbf{R}^n, A}$ such that

$$de = \sum_{i=1}^m (p_i \circ f)d(x_i \circ f).$$

Define $g : (\mathbf{R}^n, A) \to \mathbf{R}^m$ by $g(x) = (x_1 \circ f(x), \ldots, x_m \circ f(x))$. Then $e \in \mathcal{R}_g$. Conversely, given $e \in \mathcal{R}_g$, we have $de = \sum_{i=1}^m a_i dg_i$ for some functions $a_1, \ldots, a_m$, and we obtain an isotropic multi-germ $f : (\mathbf{R}^n, A) \to \mathbf{R}^{2m}$ by $p_i \circ f = a_i, x_i \circ f = g_i, (1 \leq i \leq m)$. The opening $(g, e) : (\mathbf{R}^n, A) \to \mathbf{R}^{m+1}$ is the *frontal* germ associated to f. For example the frontal of a open Whitney-Morin umbrella is the folded umbrella (see Proposition 4.1). The open folded umbrella appears also as a "frontal-symplectic singularity" [18]. Several geometric applications of the theory of openings are given in [16] [17].

Remark 1.3 (Relations with known notions). Pellikaan (his thesis and [25], p.358), de Jong and van Straten ([19], p.185) introduced the notion of *primitive ideal*

$$\int I = \{h \in \mathcal{O}_n \mid h \in I, \frac{\partial h}{\partial x_i} \in I, 1 \leq i \leq n\}$$

of an ideal I of the ring $\mathcal{O}_n$ of holomorphic function-germs on $(\mathbf{C}^n, 0)$. It is motivated for the deformation theory of non-isolated singularities. We can introduce the analogous notion to it in C^∞ case and also the notion of *primitive ring* by

$$\int' I = \{h \in \mathcal{E}_n \mid \frac{\partial h}{\partial x_i} \in I, 1 \leq i \leq n\}$$

It is related to the notion of ramification module as follows: Let $f : (\mathbf{R}^n, 0) \to (\mathbf{R}^m, 0), n \leq m$ be a map-germ. We take as I the ideal J_f generated by n-minors of the Jacobi matrix of f, which is associated with the singular locus of f. If $n = 1$ (the case of curves), then we have that

$\int' J_f = \mathcal{R}_f$. However, they are different in general in the case $n \geq 2$. In fact $f^* \mathcal{E}_m \not\subseteq \int' J_f$ and $\int' J_f$ does not have the natural $\mathcal{E}_m$-module structure via f^* in general, while $\mathcal{R}_f$ does. For example, let $f(x_1, x_2) = (x_1, x_2^2)$, $f : (\mathbf{R}^2, 0) \to (\mathbf{R}^2, 0)$. Then $\int' J_f = \mathbf{R} + x_2^2 \mathcal{E}_2$, and $\mathcal{R}_f = \mathcal{E}_1 + x_2^2 \mathcal{E}_2 = f^* \mathcal{E}_2 + x_2^3 f^* \mathcal{E}_2$ where $\mathcal{E}_1$ means the ring of functions of x_1.

Mond [24] introduced a notion $R_0(f)$, which is more closely related to $\mathcal{R}_f$, also from the motivation of deformation theory on map-germs in complex analytic category. Its C^∞ analogue is given by

$$R_0(f) = \{ h \in \mathcal{E}_n \mid df_{i_1} \wedge \cdots \wedge df_{n-1} \wedge dh \in J_f \, dx_1 \wedge \cdots \wedge dx_n,$$
$$1 \leq i_1 < \cdots < i_{n-1} \leq m \}.$$

Then clearly we have $\mathcal{R}_f \subseteq R_0(f)$. We have that, for two map-germ f, f', $\mathcal{R}_f = \mathcal{R}_{f'}$ implies $R_0(f) = R_0(f')$. If $n = 1$, then $\mathcal{R}_f = R_0(f) = \int' J_f$. Moreover we have that, if $n = m$ and the singular locus of f is nowhere dense, then $\mathcal{R}_f = R_0(f)$, using Cramer's rule in linear algebra. We conjecture that the equality $\mathcal{R}_f = R_0(f)$ holds also in the case $n < m$ under a rather mild condition. Note that, in Propositions 4.1 and 4.3 of [24], they were already given the related results to the results in the present paper (Lemma 2.2, Corollary 2.10, Proposition 2.16 and Proposition 5.2).

In §2, we give a detailed exposition on ramification modules and openings of multi-germs.

In §3, the relation of unfoldings and openings are discussed. We treat the problem to find a versal opening of an unfolding of a given map-germ. Then the notion of *extendability* of an unfolding is introduced. If the given map-germ is of corank one, then any unfolding is extendable and, in particular its versal opening is obtained from that of its stable unfolding. Then, in §4, we give the explicit presentation on versal openings of stable unfoldings of map-germs of corank 1, namely, versal openings of Morin maps.

In §5, we remark the existence of versal openings in finite analytic case. In §6, we give a direct method to find versal openings for several examples and show the existence of the versal opening for any stable map-germ $(\mathbf{R}^4, A) \to (\mathbf{R}^4, b)$, explicitly. In §7, the relation of lowerable vector fields of map-germs and openings is discussed.

In this paper we often abbreviate $\mathcal{E}_{(\mathbf{R}^n, A)}$ by $\mathcal{E}_A$, the $\mathbf{R}$-algebra of C^∞ function-germs on $(\mathbf{R}^n, A)$. If $A = \{a_1, \ldots, a_s\}$, then we denote by $\mathfrak{m}_i$ the maximal ideal consisting of $h \in \mathcal{E}_A$ with $h(a_i) = 0$. We set $\mathfrak{m}_A = \cap_{i=1}^s \mathfrak{m}_i$. If A consists of the origin, then we use $\mathcal{E}_n, \mathfrak{m}_n$ instead of $\mathcal{E}_A, \mathfrak{m}_A$ respectively.

All manifolds and mappings which we treat in this paper are assumed to be of class C^∞, unless otherwise stated.

2. Ramification modules and openings

In Introduction, we have introduced the notion of openings based on that of Jacobi modules and ramification modules.

Lemma 2.1. *For map-germs* $f : (\mathbf{R}^n, A) \to (\mathbf{R}^m, b)$, $f' : (\mathbf{R}^n, A) \to (\mathbf{R}^{m'}, b')$, *we have that* $\mathcal{J}_{f'} = \mathcal{J}_f$ *if and only if* $\mathcal{R}_{f'} = \mathcal{R}_f$.

Proof. It is clear that $\mathcal{J}_{f'} = \mathcal{J}_f$ implies $\mathcal{R}_{f'} = \mathcal{R}_f$. Conversely suppose $\mathcal{R}_{f'} = \mathcal{R}_f$. Then any component f'_j of f' belongs to $\mathcal{R}_{f'} = \mathcal{R}_f$, hence $df'_j \in \mathcal{J}_f$. Therefore $\mathcal{J}_{f'} \subseteq \mathcal{J}_f$. By the symmetry we have $\mathcal{J}_{f'} = \mathcal{J}_f$. □

Lemma 2.2. *Let* $f : (\mathbf{R}^n, A) \to (\mathbf{R}^m, b)$ *be a map-germ. Then we have*
(1) $f^* \mathcal{E}_b \subseteq \mathcal{R}_f \subseteq \mathcal{E}_A$.
(2) $\mathcal{R}_f$ *is an* $\mathcal{E}_b$-*submodule via* $f^* : \mathcal{E}_b \to \mathcal{E}_A$ *of* $\mathcal{E}_A$.
(3) $\mathcal{R}_f$ *is* C^∞-*subring of* $\mathcal{E}_A$.
(4) If $\tau : (\mathbf{R}^m, b) \to (\mathbf{R}^m, b')$ *is a diffeomorphism-germ, then* $\mathcal{R}_{\tau \circ f} = \mathcal{R}_f$. *If* $\sigma : (\mathbf{R}^n, A') \to (\mathbf{R}^n, A)$ *is a diffeomorphism-germ, then* $\mathcal{R}_{f \circ \sigma} = \sigma^*(\mathcal{R}_f)$.

Proof. Assertions (1) and (2) follow from the fact that, if $h \in \mathcal{R}_f$ and $dh = \sum_{j=1}^m p_j df_j$, then

$$d\{(k \circ f)h\} = \sum_{j=1}^m \{(k \circ f)p_j + h(\partial k/\partial y_j)\} df_j.$$

Assertion (3) follows from the fact that, if $h_1, \ldots, h_r \in \mathcal{R}_f$ and if $\tau : \mathbf{R}^r \to \mathbf{R}$ is a C^∞ function, then

$$d\{\tau(h_1, \ldots, h_r)\} = \sum_{i=1}^r \frac{\partial \tau}{\partial y_i}(h_1, \ldots, h_r) \, dh_i \in \mathcal{J}_f.$$

Assertion (4) follows from the fact that $\mathcal{J}_{\tau \circ f} = \mathcal{J}_f$ and $\mathcal{J}_{f \circ \sigma} = \sigma^*(\mathcal{J}_f)$. □

Lemma 2.3. *Let* $f : (\mathbf{R}^n, A) \to (\mathbf{R}^m, b)$ *be a map-germ with* $A = \{a_1, \ldots, a_s\}$. *We denote by* $f_i : (\mathbf{R}^n, a_i) \to (\mathbf{R}^m, b)$ *the restriction of* f *to* $(\mathbf{R}^n, a_i)$. *Then* $\mathcal{R}_f \cong \prod_{i=1}^s \mathcal{R}_{f_i}$ *as* $\mathcal{E}_b$-*module.*

Proof. We have the isomorphism $\varphi : \mathcal{R}_f \to \prod_{i=1}^s \mathcal{R}_{f_i}$ defined by $\varphi(h) = (h|_{(\mathbf{R}^n, a_i)})_{i=1}^s$. □

Remark 2.4. Any multi-germ $f : (\mathbf{R}^n, A) \to (\mathbf{R}^m, b), \#(A) = s$, is right-equivalent to a map-germ of form $\coprod_s f_i : \coprod_s (\mathbf{R}^n, 0) \to (\mathbf{R}^m, b)$ from the disjoint union of s-copies of $(\mathbf{R}^n, 0)$.

A map-germ $f : (\mathbf{R}^n, A) \to (\mathbf{R}^m, b)$ is called *finite* if $\mathcal{E}_A$ is a finite $\mathcal{E}_b$-module. The condition is equivalent to that $\dim_{\mathbf{R}} \mathcal{E}_A/(f^* \mathfrak{m}_b)\mathcal{E}_A < \infty$ by the preparation theorem (see for example [4]). Moreover f is finite if and only if $\mathcal{K}$-finite and $n \leq m$ ([26]).

Proposition 2.5. *If* $f : (\mathbf{R}^n, A) \to (\mathbf{R}^m, b)$ *is finite and of corank at most one. Then we have*
(1) $\mathcal{R}_f$ *is a finite* $\mathcal{E}_b$-*module. Therefore there exists a versal opening of* f.
(2) $1, h_1, \ldots, h_r \in \mathcal{R}_f$ *generate* $\mathcal{R}_f$ *as* $\mathcal{E}_b$-*module if and only if they generate the vector space* $\mathcal{R}_f/(f^* \mathfrak{m}_b)\mathcal{R}_f$ *over* $\mathbf{R}$.

Proof. In the case A consists of a point, the assertions are proved in Theorem 1.3 of [13] and Corollary 2.4 of [15]. For a general finite set A, the assertions are reduced to the case that A consists of a point by Lemma 2.3.
□

Remark 2.6. In §5, we define the analytic counterpart $\mathcal{R}_f^\omega$ of the notion of ramification modules for an analytic map-germ $f : (\mathbf{R}^n, A) \to (\mathbf{R}^m, 0)$. Then it is essentially obvious that, if f is a finite map-germ, then $\mathcal{R}_f^\omega$ is a finite module over the ring $\mathcal{O}_{(\mathbf{R}^n, A)}$ of analytic function-germs. In fact $\mathcal{O}_{(\mathbf{R}^n, A)}$ is a Noetherian ring and it is a finite $\mathcal{O}_{(\mathbf{R}^m, 0)}$-module via f^*. Moreover $\mathcal{R}_f^\omega$ is an $\mathcal{O}_{(\mathbf{R}^m, 0)}$-submodule of $\mathcal{O}_{(\mathbf{R}^n, A)}$. Since every submodule of a finite module over a Noetherian ring is finite, we have that $\mathcal{R}_f^\omega$ is a finite module over the ring $\mathcal{O}_{(\mathbf{R}^n, A)}$.

Furthermore, in §5, we show that $\mathcal{R}_f$ is a finite $\mathcal{E}_b$-module if f is finite and analytic (Proposition 5.2).

For a map-germ $f : (\mathbf{R}^n, A) \to (\mathbf{R}^m, b), n \leq m$, we have defined in Introduction the notions of openings and versal openings of f.

Example 2.7. (1) Let $h : (\mathbf{R}, 0) \to (\mathbf{R}, 0)$, $h(x) = x^2$. Then $\mathcal{R}_h = \langle 1, x^3 \rangle_{f^*(\mathcal{E}_1)}$. The map-germ $H : (\mathbf{R}, 0) \to (\mathbf{R}^2, 0), H(x) := (x^2, x^3)$, the simple cusp map, is the mini-versal opening of h.
(2) Let $g : (\mathbf{R}, 0) \to (\mathbf{R}, 0)$, $g(x) = x^3$. Then $\mathcal{R}_g = \langle 1, x^4, x^5 \rangle_{f^*(\mathcal{E}_1)}$. The map-germ $G : (\mathbf{R}, 0) \to (\mathbf{R}^3, 0)$, $G(x) := (x^3, x^4, x^5)$ is the mini-versal opening of g.

(3) Let $f : (\mathbf{R}^2, 0) \to (\mathbf{R}^2, 0)$, $f(x, u) = (x^3 + ux, \ u)$, an unfolding of g. Then $\mathcal{R}_f = \langle 1, \ x^4 + \frac{2}{23}ux^2, \ x^5 + \frac{4}{9}ux^3 \rangle_{f^*(\mathcal{E}_2)}$. The map-germ $F : (\mathbf{R}^2, 0) \to (\mathbf{R}^4, 0)$ defined by $F(x, u) := (x^3 + ux, \ u, \ x^4 + \frac{2}{23}ux^2, \ x^5 + \frac{4}{9}ux^3)$, the open swallowtail is the mini-versal opening of Whitney cusp f.

(4) Let consider the multi-germ $k : (\mathbf{R}, A) \to (\mathbf{R}^2, 0)$, $A = \{0, 1\}$ defined by $k(t) = (t, 0)$ near $t = 0$ and $k(t) = (0, t - 1)$ near $t = 1$. Then $\mathcal{R}_k$ is generated by $1 \in \mathcal{E}_A$ and $s \in \mathcal{E}_A$ defined by $s(t) = 1$ near $t = 0$ and $s(t) = 0$ near $t = 1$, over $k^*\mathcal{E}_2$. Then $K = (k, s) : (\mathbf{R}, A) \to \mathbf{R}^3$ is the versal opening of k. In fact K resolves the self-intersection by the over and underpasses.

Example 2.8. Here we add several additional illustrative examples. Let us consider the following five map-germs: $f : (\mathbf{R}^2, 0) \to (\mathbf{R}^2, 0)$, $g : (\mathbf{R}^2, 0) \to (\mathbf{R}^3, 0)$, $h : (\mathbf{R}^2, 0) \to (\mathbf{R}^3, 0)$, $k : (\mathbf{R}^2, 0) \to (\mathbf{R}^2, 0)$, $\ell : (\mathbf{R}^2, 0) \to (\mathbf{R}^2, 0)$ defined by

$$f(x, t) = (x, \ t^2), \quad g(x, t) = (x, \ xt, \ t^2), \quad h(x, t) = (x^2, \ xt, \ t^2),$$

$$k(x, t) = (x^2, \ t^2), \quad \ell(x, t) = (x^2 - t^2, \ xt).$$

Then we have

$$\mathcal{R}_k \subsetneqq \mathcal{R}_h \subsetneqq \mathcal{R}_g, \quad \mathcal{R}_\ell \subsetneqq \mathcal{R}_h, \quad \mathcal{R}_f \subsetneqq \mathcal{R}_g.$$

In fact

$$\mathcal{J}_f = \langle dx, t \, dt \rangle_{\mathcal{E}_2}, \quad \mathcal{J}_g = \langle dx, x \, dt, t \, dt \rangle_{\mathcal{E}_2}, \quad \mathcal{J}_h = \langle x \, dx, x \, dt + t \, dx, t \, dt \rangle_{\mathcal{E}_2},$$

$$\mathcal{J}_k = \langle x \, dx, t \, dt \rangle_{\mathcal{E}_2}, \quad \mathcal{J}_\ell = \langle x \, dx - t \, dt, t \, dx + x \, dt \rangle_{\mathcal{E}_2}.$$

Then we see that $\mathcal{R}_f$ is minimally generated by $1, t^3$ over $f^*\mathcal{E}_2$, $\mathcal{R}_g$ is minimally generated by $1, t^3$ over $g^*\mathcal{E}_3$, and $\mathcal{R}_h$ is minimally generated by $1, x^3, x^2t, xt^2, t^3$ over $h^*\mathcal{E}_3$. Moreover we have that $\mathcal{R}_k$ is minimally generated by $1, x^3, t^3, x^3t^3$ over $k^*\mathcal{E}_2$ and that $\mathcal{R}_\ell$ is minimally generated by $1, x^3 - 3xt^2, 3x^2t - t^3, x^2(x^2 + t^2)^2$ over $\ell^*\mathcal{E}_2$.

Remark 2.9 (continued with Remark 2.4). Let $\coprod_s f_i : \coprod_s(\mathbf{R}^n, 0) \to (\mathbf{R}^m, b)$ be a multi-germ of map. Suppose $F_i = (f_i; h_{i1}, \ldots, h_{ir_i}) : (\mathbf{R}^n, 0) \to (\mathbf{R}^{m+r_i}, b \times 0)$ be a versal opening of f_i. Then, setting $r = \max_{1 \le i \le s} r_i$ and $h_{ij} = 0$ if $j > r_i$, then $F = \coprod_s(f_i, h_{i1}, \ldots, h_{ir_i}, \ldots) : \coprod_s(\mathbf{R}^n, 0) \to (\mathbf{R}^{m+r}, b \times 0)$ is a versal opening of $\coprod_s f_i$.

By Proposition 2.5, we have

Corollary 2.10. Let $f : (\mathbf{R}^n, A) \to (\mathbf{R}^m, b)$ be finite and of corank at most one. Then there exists a versal opening of f.

Remark 2.11. The existence of a versal opening for a finite C^∞ map-germ of arbitrary corank is still open. However, using Proposition 5.2 together with the analytic result, we can show that Corollary 2.10 for finitely $\mathcal{A}$-determined map-germs $(\mathbf{R}^n, A) \to (\mathbf{R}^m, b)$ with $n \le m$ without the corank condition.

Moreover we have the following:

Corollary 2.12. *Let $f : (\mathbf{R}^n, A) \to (\mathbf{R}^m, b)$ be finite and of corank at most one. Then an opening $F = (f, h_1, \ldots, h_r)$ of f is a mini-versal opening of f, namely, $1, h_1, \ldots, h_r \in \mathcal{R}_f$ form a minimal system of generators of $\mathcal{R}_f$ as $\mathcal{E}_b$-module if and only if they form a basis of $\mathbf{R}$-vector space $\mathcal{R}_f / (f^* \mathfrak{m}_b) \mathcal{R}_f$.*

The following is useful for the classification problem of map-germs in a geometric context ([16,17]).

Proposition 2.13. *Let $f : (\mathbf{R}^n, A) \to (\mathbf{R}^m, b)$, $n \le m$ be a C^∞ map-germ.*
(1) For any versal opening $F : (\mathbf{R}^n, A) \to (\mathbf{R}^{m+r}, F(A))$ of f and for any opening $G : (\mathbf{R}^n, A) \to (\mathbf{R}^{m+s}, G(A))$, there exists an affine bundle map $\Psi : (\mathbf{R}^{m+r}, F(A)) \to (\mathbf{R}^{m+s}, G(A))$ over $(\mathbf{R}^m, b)$ such that $G = \Psi \circ F$.
(2) For any mini-versal openings $F : (\mathbf{R}^n, A) \to (\mathbf{R}^{m+r}, F(A))$ and $F' : (\mathbf{R}^n, A) \to (\mathbf{R}^{m+r}, F'(A))$ of f, there exists an affine bundle iso-morphism $\Phi : (\mathbf{R}^{m+r}, F(A)) \to (\mathbf{R}^{m+r}, F'(A))$ over $(\mathbf{R}^m, b)$ such that $F' = \Psi \circ F$. In particular, the diffeomorphism class of mini-versal opening of f is unique.
(3) Any versal openings $F'' : (\mathbf{R}^n, A) \to (\mathbf{R}^{m+s}, F''(A))$ of f is diffeomor-phic to $(F, 0)$ for a mini-versal opening F of f.

Proof. (1) Let $F = (f, h_1, \ldots, h_r)$ and $G = (f, k_1, \ldots, k_s)$. Since $k_j \in \mathcal{R}_f$, there exist $c_j{}^0, c_j{}^1, \ldots, c_j{}^r \in \mathcal{E}_b$ such that $k_j = c_j{}^0 \circ f + (c_j{}^1 \circ f) h_1 + \cdots + (c_j{}^r \circ f) h_r$. Then it suffices to set

$$\Psi(y, z) = (y, (c_j{}^0(y) + c_j{}^1(y) z_1 + \cdots + c_j{}^r(y) z_r)_{1 \le j \le s}).$$

(2) By (1) there exists an affine bundle map Ψ with $F' = \Psi \circ F$. From the minimality, we have that the matrix $(c_j{}^i(b))$ is regular. (See Corollary 2.12.) Therefore Ψ is a diffeomorphism-germ.
(3) Let $F = \Psi \circ F''$ for some affine bundle map Ψ. Then the matrix $(c_j{}^i(b))$ is of rank r. Therefore F'' is diffeomorphic to $(F, k_1, \ldots, k_{s-r})$ for some $k_j \in \mathcal{R}_f$. Write each $k_j = K_j \circ F$ for some $K_j \in \mathcal{E}_{F(a)}$. Then we set $\Xi(y, z, w) = (y, z, w - K \circ F)$. Then Ξ is a local diffeomorphism on $\mathbf{R}^{m+r+(s-r)}$ and $\Xi \circ (F, k_1, \ldots, k_{s-r}) = (F, 0)$. $\square$

Two map-germs $F : (\mathbf{R}^n, A) \to (\mathbf{R}^p, B)$ and $G : (\mathbf{R}^n, A) \to (\mathbf{R}^q, C)$ is called $\mathcal{L}$-equivalent, or, left-equivalent, if there exists a diffeomorphism-germ $\Psi : (\mathbf{R}^p, B) \to (\mathbf{R}^q, C)$ such that $G = \Psi \circ F$.

Then, by Proposition 2.13, we have:

Corollary 2.14. *Let $f : (\mathbf{R}^n, A) \to (\mathbf{R}^m, b)$ be a C^∞ map-germ $(n \le m)$. Then a mini-versal opening of f is unique up to $\mathcal{L}$-equivalence. A versal opening of f is $\mathcal{L}$-equivalent to a mini-versal opening composed with an immersion.*

We introduce further the following notion:

Definition 2.15. An opening $F = (f, h_1, \ldots, h_s)$ of a map-germ

$$f = (f_1, \ldots, f_m) : (\mathbf{R}^n, A) \to (\mathbf{R}^m, b)$$

is called $\mathcal{L}$-*minimal* if $f_1, \ldots, f_m, h_1, \ldots, h_r$ minimally generate the C^∞-ring $\mathcal{R}_f$ over $\mathbf{R}$, in other words, if $\mathcal{R}_f = (f, h)^*(\mathcal{E}_{m+r})$ and (f, h) in minimal with this property.

Then we have a similar uniqueness result for $\mathcal{L}$-equivalence of $\mathcal{L}$-minimal openings.

In Example 2.8, we have seen that $1, x^3, t^3, x^3 t^3$ minimally generate $\mathcal{R}_k$ as $k^* \mathcal{E}_2$-module. However $1, x^3, t^3$ already minimally generate $\mathcal{R}_k$ as $k^* \mathcal{E}_2$-C^∞-ring. Therefore $K = (x^2, y^2, x^3, t^3, x^3 t^3)$ is a mini-versal opening of $k = (x^2, y^2) : (\mathbf{R}^2, 0) \to (\mathbf{R}^2, 0)$ and $K' = (x^2, y^2, x^3, t^3)$ is a $\mathcal{L}$-minimal opening of k.

Lastly we show injectivity of versal openings:

Proposition 2.16. *Let $f : (\mathbf{R}^n, A) \to (\mathbf{R}^m, b)$ be a finite map-germ. Suppose $F : (\mathbf{R}^n, A) \to (\mathbf{R}^{m+r}, F(A))$ is a versal opening of f. Then F has an injective representative.*

Remark 2.17. The corresponding result to Proposition 2.16 in the analytic case is already proved in an earlier paper (Corollary 1.2 of [13]). By Proposition 5.2 proved in §5 together with the analytic result, we can show that Proposition 2.16 for finitely $\mathcal{A}$-determined map-germs $(\mathbf{R}^n, A) \to (\mathbf{R}^m, b)$ with $n \le m$. Note that the proof of Corollary 1.2 in [13] heavily depends, via the finite coherence theorem in analytic geometry, on that the map-germ is finite and analytic. It seems very difficult to find a unified proof of Proposition 2.16 for C^∞ case with a similar vein to the analytic case, because of the lack of the notion of "coherence" in the general C^∞ case. Here we provide an alternative proof applicable to any finite C^∞ map-germs, which can be applied to finite real analytic map-germs, of course. The proof is similar to the proof for a similar result which was given in [9].

Proof of Proposition 2.16. Let $A = \{a_1, \ldots a_s\}$. If each restriction $F|_{(\mathbf{R}^n, a_i)}$ for $i = 1, 2, \ldots, s$ is injective, then F is injective. In fact, for each i and for each $v \in \mathbf{R}^m$, we can take a map-germ $F^{i,v} : (\mathbf{R}^n, A) \to (\mathbf{R}^m, F_i(A))$ which coincides with F near $a_j, (j \neq i)$ and $F + v$ near a_i. Since any $F^{i,v}$ is an opening of f, and F is a versal opening, different branches of F can not intersect each other. (See Proposition 2.13 (1)).

Now suppose that f is a mono-germ $f : (\mathbf{R}^n, a) \to (\mathbf{R}^m, b)$. Assume the versal opening map-germ $F = (f, h) : (\mathbf{R}^n, a) \to (\mathbf{R}^{m+r}, F(a))$ of f has no injective representative. Then there must be a sequence of points b_i in $\mathbf{R}^{m+r}$ which tends to $F(a)$ when $i \to \infty$, and $a_i^1, a_i^2, a_i^1 \neq a_i^2$ in $\mathbf{R}^n$ which tend to a respectively when $i \to \infty$, such that $F(a_i^1) = F(a_i^2) = b_i$. We may suppose $b_i \neq b_j$ if $i \neq j$. Take a C^∞ function h on $\mathbf{R}^m$ such that the support is a disjoint union of small balls centred at b_i, $i = 1, 2, \ldots$. Note that such function must be infinitely flat at b, that is, $j^\infty h(b) = 0$. Take the C^∞ function $k = F^* h$ on $\mathbf{R}^n$. Since F is finite, the support of k is a disjoint union of closed neighbourhoods of a_i^1 and those of a_i^2, after shrinking the neighbourhood of $\{b_i\}$ on which h is non-zero if necessary. Take the function k' on $\mathbf{R}^n$ which coincides with k except on the closed neighbourhoods of a_i^2, and is identically zero there. Then we see that k' is C^∞, k' belongs to $\mathcal{R}_F = \mathcal{R}_f$ and $k'(a_i^1) \neq k'(a_i^2)$ for any $i = 1, 2, \ldots$.

In fact, k' is C^∞ on $\mathbf{R}^n \setminus \{0\}$ and it extends to an infinitely flat function on $\mathbf{R}^n$ at a. Moreover we have $dk = \sum_{i=1}^m F^*(\partial h/\partial y_i) df_i + \sum_{j=1}^r F^*(\partial h/\partial z_j) dh_j$, where $y_1, \ldots, y_m, z_1, \ldots, z_r$ are the coordinates of $\mathbf{R}^{m+r}$. Take the function a_i (resp. b_j) on $\mathbf{R}^n$ which coincides with $F^*(\partial h/\partial y_i)$ (resp. $F^*(\partial h/\partial z_j)$) except on the closed neighbourhoods of a_i^2, and is identically zero there. Then a_i and b_j are C^∞ and, then we have $dk' = \sum_{i=1}^m a_i df_i + \sum_{j=1}^r b_j dh_j \in \mathcal{J}_F = \mathcal{J}_f$. Therefore we have $k' \in \mathcal{R}_f$.

Consider the opening $(f, k') : (\mathbf{R}^n, a) \to (\mathbf{R}^{m+1}, (b, 0))$ of f. Since F is a versal opening of f, we must have $k' = \tau \circ F$ for a function-germ $\tau : (\mathbf{R}^n, a) \to \mathbf{R}$. See Proposition 2.13 (1), or, the definition of versal openings (Definition 1.2). Then we must have $k'(a_i^1) = \tau(F(a_i^1)) = \tau(F(a_i^2)) = k'(a_i^2)$ for a sufficiently large i. This leads a contradiction. Therefore we have that F is injective. $\qquad\square$

Remark 2.18. By Proposition 2.16, the versal opening $F = (f, h_1, \ldots, h_r)$ of a finite map-germ f is injective. However F is not necessarily finitely $\mathcal{A}$-determined. For example, the cuspidal edge is a versal opening of the fold map and it is not finitely $\mathcal{A}$-determined (Example 2.8 and Remark 2.18). In fact a higher order perturbation of F destroys the singular locus of F. On the other hand, whether an opening F of f is versal or not depends only on finite jets of $h_1, \ldots, h_r$. This follows from the preparation theorem (cf. Proposition 2.13).

Remark 2.19. Related to the above Proposition 2.16, T. Gaffney suggested a relation of the notion of openings and that of weak normalisations [6], which is left to be our open problem.

3. Unfoldings and openings

We recall the notion of unfolding of map-germs ([21]).

Let $f : (\mathbf{R}^n, A) \to (\mathbf{R}^m, b)$ be a map-germ. An unfolding of f is a map-germ $F : (\mathbf{R}^{n+\ell}, A \times 0) \to (\mathbf{R}^{m+\ell}, (b, 0))$ of form $F(x, u) = (F_1(x, u), u)$ and $F_1(x, 0) = f(x)$, for $(x, u) \in (\mathbf{R}^{n+\ell}, A \times 0)$.

For another unfolding $G : (\mathbf{R}^{n+\ell}, A \times 0) \to (\mathbf{R}^{m+\ell}, (b, 0))$, F and G are called *isomorphic* if there exist an unfolding $\Sigma : (\mathbf{R}^{n+\ell}, A \times 0) \to (\mathbf{R}^{n+\ell}, A \times 0)$ of the identity map on $(\mathbf{R}^n, A)$ and an unfolding $T : (\mathbf{R}^{m+\ell}, (b, 0)) \to (\mathbf{R}^{m+\ell}, (b, 0))$ of the identity map on $(\mathbf{R}^m, b)$ such that $G \circ \Sigma = T \circ F$.

Proposition 3.1 (Unfoldings and openings). *Let* $f : (\mathbf{R}^n, A) \to (\mathbf{R}^m, b)$ *be a* C^∞ *map-germ and* $F : (\mathbf{R}^{n+\ell}, A \times 0) \to (\mathbf{R}^{m+\ell}, (b, 0))$ *be an unfolding of* f. *Let* $i : (\mathbf{R}^n, A) \to (\mathbf{R}^{n+\ell}, A \times 0)$ *be the inclusion,* $i(x) = (x, 0)$. *Then we have:*

(1) $i^* \mathcal{R}_F \subset \mathcal{R}_f$.

(2) If f *is of corank* ≤ 1 *with* $n \leq m$, *then* $i^* \mathcal{R}_F = \mathcal{R}_f$. *If* $1, H_1, \ldots, H_r$ *generate* $\mathcal{R}_F$ *via* F^*, *then* $1, i^* H_1, \ldots, i^* H_r$ *generate* $\mathcal{R}_f$ *via* f^*.

Proof. For the mono-germ case the assertions are proved in Proposition 1.6 of [13], Lemma 2.4 of [14]. Here we present the proof for the general case: (1) is clear. (2) Let $H \in \mathcal{R}_F$. Then $dH \in \mathcal{J}_F$. Hence $d(i^* H) = i^*(dH) \in i^* \mathcal{J}_F \subset \mathcal{J}_f$. Therefore $i^* H \in \mathcal{R}_f$. Let f be of corank at most one. Suppose $h \in \mathcal{R}_f$. Then $dh = \sum_{j=1}^m a_j df_j$ for some $a_j \in \mathcal{E}_a$. There exist $A_j, B_k \in \mathcal{E}_{(a,0)}$ such that $i^* A_j = a_j$ and the 1-form $\sum_{j=1}^m A_j d(F_1)_j + \sum_{k=1}^\ell B_k d\lambda_k$ is closed (cf. Lemma 2.5 of [15]). Then there exists an $H \in \mathcal{E}_{(a,0)}$ such that $dH = \sum_{j=1}^m A_j d(F_1)_j + \sum_{k=1}^\ell B_k d\lambda_k \in \mathcal{J}_F$ and $d(i^* H) = i^*(dH) = dh$. Then there exists $c \in \mathbf{R}$ such that $h = i^*H + c = i^*(H + c)$, and $H + c \in \mathcal{R}_F$. Therefore $h \in i^*\mathcal{R}_F$. Since i^* is a homomorphism over $j^* : \mathcal{E}_{(b,0)} \to \mathcal{E}_b$, where $j : (\mathbf{R}^m, 0) \to (\mathbf{R}^{m+\ell}, 0)$ is the inclusion $j(y) = (y, 0)$, we have the consequence. $\qquad\square$

An unfolding $F : (\mathbf{R}^{n+\ell}, A \times 0) \to (\mathbf{R}^{m+\ell}, (b, 0))$ of a map-germ $f : (\mathbf{R}^n, A) \to (\mathbf{R}^m, b)$ is called *extendable* if $i^*\mathcal{R}_F = \mathcal{R}_f$ for the inclusion $i : (\mathbf{R}^n, A) \to (\mathbf{R}^{n+\ell}, A \times 0)$. By Proposition 3.1, we have:

Corollary 3.2. *If corank of f is at most one, then any unfolding of f is extendable.*

In §6, we will see that there exist non-extendable unfoldings for map-germs of corank ≥ 2. Therefore the opening constructions do not behave well under unfoldings in general.

4. Openings of stable maps of corank one

We will give the explicit versal opening in the case of corank one. As is seen in Remark 2.9, it is sufficient to treat the case of mono-germs, namely, germs $f : (\mathbf{R}^n, 0) \to (\mathbf{R}^m, 0)$ of corank one. Moreover, by Corollary 3.2, it is sufficient to treat the case that f is stable, namely, f is a Morin map.

Let $k \geq 0, m \geq 0$. To present the normal forms of Morin maps, consider variables $t, \lambda = (\lambda_1, \ldots, \lambda_{k-1}), \mu = (\mu_{ij})_{1 \leq i \leq m, 1 \leq j \leq k}$ and polynomials

$$F(t, \lambda) = t^{k+1} + \sum_{i=1}^{k-1} \lambda_j t^j, \quad G_i(t, \mu) = \sum_{j=1}^{k} \mu_{ij} t^j, (1 \leq i \leq m).$$

Let $f : (\mathbf{R}^{k+km}, 0) \to (\mathbf{R}^{m+k+km}, 0)$ be a Morin map defined by

$$f(t, \lambda, \mu) := (F(t, \lambda), G(t, \mu), \lambda, \mu),$$

for the above polynomials F and G.

For $\ell \geq 0$, we denote by $F_{(\ell)}, G_{i(\ell)}$ the polynomials

$$F_{(\ell)}(t, \lambda) = \int_0^t s^\ell F(s, \lambda) ds, \quad G_{i(\ell)}(t, \mu) = \int_0^t s^\ell G_i(s, \mu) ds.$$

Then we have:

Proposition 4.1 (Theorem 3 of [12]). *The ramification module $\mathcal{R}_f$ of the Morin map f is minimally generated over $f^*\mathcal{E}_{m+k+km}$ by the $1 + k + (k-1)m$ elements*

$$1, \ F_{(1)}, \ \ldots, \ F_{(k)}, \ G_{1(1)}, \ \ldots, \ G_{1(k-1)}, \ \ldots, \ G_{m(1)}, \ \ldots, \ G_{m(k-1)}.$$

The map-germ

$$\mathbf{F} : (\mathbf{R}^{k+mk}, 0) \to (\mathbf{R}^{m+k+km} \times \mathbf{R}^{k+(k-1)m}, 0) = (\mathbf{R}^{2(k+km)}, 0)$$

defined by

$$\mathbf{F} = \left(f,\ F_{(1)}, \ldots, F_{(k)}, G_{1\,(1)}, \ldots, G_{1\,(k-1)}, \ldots, G_{m\,(1)}, \ldots, G_{m\,(k-1)} \right)$$

is a mini-versal opening of f.

Proof. The first half is proved in [12]. The second half follows from the definition. $\square$

Remark 4.2. It is shown in [12] moreover that $\mathbf{F}$ is an isotropic map for a symplectic structure on $\mathbf{R}^{2(k+km)}$. Note that the fact that the open swallowtail is a Lagrangian variety with respect to a certain symplectic structure was found first by Arnol'd and Givental' (see [8]).

In particular we have:

Lemma 4.3. *Let ℓ be a positive integer and $F = (F_1(t, u), u) : (\mathbf{R}^n, 0) \to (\mathbf{R}^n, 0)$ an unfolding of $f : (\mathbf{R}, 0) \to (\mathbf{R}, 0)$, $f(t) = F_1(t, 0) = t^\ell$. Suppose $H_1, \ldots, H_r \in \mathcal{R}_F \cap \mathfrak{m}_n$. Then $1, H_1, \ldots, H_r$ generate $\mathcal{R}_F$ via F^* if and only if $i^* H_1, \ldots, i^* H_r$ generate $\mathfrak{m}_1^{\ell+1} / \mathfrak{m}_1^{2\ell}$. In particular $F_{1(1)}, \ldots, F_{1(\ell-1)}$ form a system of generators of $\mathcal{R}_F$ via F^* over $\mathcal{E}_n$.*

Proof. It is easy to show that $\mathcal{R}_f = \mathbf{R} + \mathfrak{m}_1^\ell$. By Proposition 2.5 (2), $1, H_1, \ldots, H_r$ generate $\mathcal{R}_F$ as $\mathcal{E}_n$-module via F^* if and only if they generate $\mathcal{R}_F / F^*(\mathfrak{m}_n) \mathcal{R}_F$ over $\mathbf{R}$. Since

$$\mathcal{R}_F / F^*(\mathfrak{m}_n) \mathcal{R}_F \cong (\mathbf{R} + \mathfrak{m}_1^\ell) / (f^* \mathfrak{m}_1)(\mathbf{R} + \mathfrak{m}_1^\ell) \cong \mathfrak{m}_1^{\ell+1} / \mathfrak{m}_1^{2\ell}$$

we have the consequence. $\square$

5. Versal openings of analytic map-germs

In this section we discuss the case f is analytic.

First we recall the complex analytic case briefly from [13].

Let (X, A) be a germ of complex analytic space at a finite set A with the structure sheaf $\mathcal{O}_{X,A}$, and $f = (f_1, \ldots, f_m) : (X, A) \to (\mathbf{C}^m, b)$ a finite analytic map-germ. In the graded differential $\mathcal{O}_{X,A}$-algebra (de Rham algebra) $\Omega_{X,A}$ on (X, A), consider the graded differential ideal $\mathcal{I}_f$ generated by $df_1, \ldots, df_m$. Then the differential d on $\Omega_{X,A}$ induces the $f^* \mathcal{O}_{\mathbf{C}^m}$-homomorphism $d : \Omega_{X,A} / \mathcal{I}_f \to \Omega_{X,A} / \mathcal{I}_f$ and then $\mathcal{O}_{\mathbf{C}^m, b}$-homomorphism

$d : f_*(\Omega_{X,A}/\mathcal{I}_f) \to f_*(\Omega_{X,A}/\mathcal{I}_f)$. Then we consider the i-th cohomology $\mathcal{H}^i(f_*(\Omega_{X,A}/\mathcal{I}_f); d)$ for the complex $(f_*(\Omega_{X,A}/\mathcal{I}_f), d)$. It is evident that $\mathcal{H}^i(f_*(\Omega_{X,A}/\mathcal{I}_f); d)$ is a finite $\mathcal{O}_{\mathbf{C}^m,b}$-module (cf. Remark 2.6). In fact, we have moreover:

Proposition 5.1 (Proposition 1.1 of [13]). $\mathcal{H}^i(f_*(\Omega_{X,A}/\mathcal{I}_f); d)$ is a coherent $\mathcal{O}_{\mathbf{C}^m,b}$-module $(i = 0, 1, 2, \dots)$.

We remark that the stalk of $\mathcal{H}^0(f_*(\Omega_{X,A}/\mathcal{I}_f); d)$ at A is the complex analytic counterpart of $\mathcal{R}_f$ in the real C^∞ case. We write it $\mathcal{R}_f^{\mathrm{hol}}$ to distinguish with the real C^∞ case. By Proposition 5.1, in particular, that $\mathcal{R}_f^{\mathrm{hol}}$ is a finite $\mathcal{O}_{\mathbf{C}^m}$-module.

Now let $f : (\mathbf{R}^n, A) \to (\mathbf{R}^m, b)$ be a finite real analytic map-germ. We denote by $\mathcal{O}_{\mathbf{R}^n,A}$ (resp. $\mathcal{O}_{\mathbf{R}^m,b}$) the germ of sheaf of analytic functions on $(\mathbf{R}^n, A)$ (resp. $(\mathbf{R}^m, b)$). Then, besides with $\mathcal{R}_f$, we consider the sheaf

$$\mathcal{R}_f^\omega := \{h \in \mathcal{O}_{\mathbf{R}^n,A} \mid dh \in \langle df_1, \dots, df_m \rangle_{\mathcal{O}_{\mathbf{R}^n,A}}\}$$

and the direct image $f_*(\mathcal{R}_f^\omega)$ as $\mathcal{O}_{\mathbf{R}^m,b}$-module. Then we see that, in particular, $f_*(\mathcal{R}_f^\omega)$ is a finite $\mathcal{O}_{\mathbf{R}^m,b}$-module by Proposition 5.1 in the case $X = (\mathbf{C}^n, A)$. Thus we have that $f_*(\mathcal{R}_f^\omega)$ is generated over $\mathcal{O}_{\mathbf{R}^m,b}$ by some $1, h_1, \dots, h_r \in \mathcal{R}_f^\omega$. Moreover it turns out that $F = (f, h_1, \dots, h_r) : (\mathbf{R}^n, A) \to (\mathbf{R}^{m+r}, b \times h(A))$ is injective (See [13]).

Then we show the following:

Proposition 5.2. Let $f : (\mathbf{R}^n, 0) \to (\mathbf{R}^m, 0)$ be a finite analytic map-germ. Suppose $1, h_1, \dots, h_r$ generate $\mathcal{R}_f^\omega$ over $\mathcal{O}_{\mathbf{R}^m,0}$ via f^*. Then $1, h_1, \dots, h_r$ generate $\mathcal{R}_f$ over $\mathcal{E}_{\mathbf{R}^m,0}$ via f^*.

Proof. First we may suppose $h_i(0) = 0, (1 \le i \le r)$. The opening

$$F = (f_1, \dots, f_m, h_1, \dots, h_r) : (\mathbf{R}^n, 0) \to (\mathbf{R}^{m+r}, 0)$$

of f is injective by [13].

Let $\mathcal{F}_p$ stand for the $\mathbf{R}$-algebra of formal functions on $(\mathbf{R}^n, p)$. $(\mathcal{F}_p = \mathcal{E}_{\mathbf{R}^n,p}/\mathfrak{m}_{\mathbf{R}^n,p}^\infty$, and it is the completion of $\mathcal{O}_{\mathbf{R}^n,p}$ as well.) Define a sheaf $\widehat{\mathcal{F}}_{\mathbf{R}^n}$ on $\mathbf{R}^n$ by $\widehat{\mathcal{F}}_{\mathbf{R}^n}(U) = \prod_{p \in U} \mathcal{F}_p$, for any open subset U of $\mathbf{R}^n$. Denote by $\widetilde{\mathcal{F}}_n$ the stalk of $\widehat{\mathcal{F}}_{\mathbf{R}^n}$ at 0. For the definition see also Ch. III §4, pp. 45–46 of [20]. Then $\widetilde{\mathcal{F}}_n$ is faithfully flat over $\mathcal{O}_n$ (Ch. III §4, p. 47, Corollary 4.13 of [20]). Define the formal counterpart

$$\widetilde{\mathcal{R}}_f = \{(\widehat{g}_p)_{p \in (\mathbf{R}^n, 0)} \in \widetilde{\mathcal{F}}_n \mid d\widehat{g}_p \in \langle d\widehat{f}_{1,p}, \dots, d\widehat{f}_{n,p} \rangle_{\mathcal{F}_p}, \ p \in (\mathbf{R}^n, 0)\}$$

of $\mathcal{R}_f$. Here $\widehat{g}_p$ means the formal function at a point p near 0 defined by $g \in \mathcal{E}_{\mathbf{R}^n, p}$. Then, for our $F = (f, h_1, \ldots, h_r)$, we have that $1, h_1, \ldots, h_r$ generate $\widetilde{\mathcal{R}}_f$ over $\widetilde{F}_m$.

For a map-germ $F : (\mathbf{R}^n, 0) \to (\mathbf{R}^{m+r}, 0)$, we say that a function-germ $g \in \mathcal{E}_n$ *formally belongs* to $F^* \mathcal{E}_{m+r}$ if, for any $q \in (\mathbf{R}^{m+r}, 0)$, there exists a formal function $\widehat{k} \in \mathcal{F}_q$ such that, for any $p \in F^{-1}(q)$, $\widehat{g}_p = \widehat{F^* k}$. Then, it is known the following Glaeser's type theorem on characterisation of composite differentiable functions: If $g \in \mathcal{E}_n$ formally belongs to $F^* \mathcal{E}_{m+r}$, then g belongs to $F^*(\mathcal{E}_{m+r})$. In fact the theorem follows, for instance, from Theorem D and Theorem C (3) of [2]. (We apply Theorem D of [2] to the case that $\phi = F$ is a finite map-germ, and $X = (\mathbf{R}^n, 0)$, $Y = (\mathbf{R}^{m+r}, 0)$, $s = 1$, $p = q = r = 1$, $A = 1$, $B = 0$ for the notations in [2].) See also Theorem 11.8 of [3].

Since F is injective in our case, we have that any element of $\mathcal{R}_f$ formally belongs to $F^* \mathcal{E}_{m+r}$, therefore we have $\mathcal{R}_f \subseteq F^* \mathcal{E}_{m+r}$. Since F is an opening of f, we have that $\mathcal{R}_f = F^* \mathcal{E}_{m+r}$.

Let $\pi : (\mathbf{R}^{m+r}, 0) \to (\mathbf{R}^m, 0)$ be the projection. Then $\pi^* : \mathcal{E}_m \to \mathcal{E}_{m+r}$ is the inclusion. Regard $F^* \mathcal{E}_{m+r}$ as an $\mathcal{E}_{m+r}$-module via F^*. By the preparation theorem, $1, h_1, \ldots, h_r$ generate $F^* \mathcal{E}_{m+r} = \mathcal{R}_f$ as $\mathcal{E}_m$-module via $F^* \circ \pi^* = f^*$ if $1, h_1, \ldots, h_r$ generate $F^* \mathcal{E}_{m+r} / (f^* \mathfrak{m}_m) F^* \mathcal{E}_{m+r}$ over $\mathbf{R}$.

We will show that

$$\mathfrak{m}_n^\infty \cap F^* \mathcal{E}_{m+r} \subseteq (f^* \mathfrak{m}_m) F^* \mathcal{E}_{m+r}.$$

Define $h = \sum_{i=1}^m f_i^2 : (\mathbf{R}^n, 0) \to (\mathbf{R}, 0)$. Since f is finite, $h^{-1}(0) = \{0\}$, and moreover, the norms of $1/h$ and its partial derivatives up to order say ℓ are bounded above by $1/\|x\|^\alpha$ for some $\alpha = \alpha(\ell) > 0$. Then $1/h$ is a multiplier for the ideal $\mathfrak{m}_n^\infty$ in the sense of Malgrange (Ch. IV §1, p.54, Proposition 1.4 of [20]). Hence, for any $k \in \mathfrak{m}_n^\infty$, k/h is a C^∞ function on $(\mathbf{R}^n, 0)$ and it is an element of $\mathfrak{m}_n^\infty$, if we set $(k/h)(0) = 0$. Moreover let $k \in \mathfrak{m}_n^\infty \cap F^* \mathcal{E}_{m+r}$. Then k/h formally belongs to $F^* \mathcal{E}_{m+r}$. In fact, the Taylor series of k/h at $0 \in \mathbf{R}^n$ is 0, and, outside of 0, k/h is a composite function of F. Therefore, again by the Glaeser's type theorem as above, we have $k/h \in F^* \mathcal{E}_{m+r}$. Then

$$k = (\sum_{i=1}^m f_i^2)(k/h) = \sum_{i=1}^m f_i(f_i k/h) \in (f^* \mathfrak{m}_m) F^* \mathcal{E}_{m+r}.$$

Thus we have $\mathfrak{m}_n^\infty \cap F^* \mathcal{E}_{m+r} \subseteq (f^* \mathfrak{m}_m) F^* \mathcal{E}_{m+r}$.

Now let $H \in \mathcal{R}_f = F^* \mathcal{E}_{m+r}$. Then

$$H \equiv a_0 \circ f + a_1 \circ f \cdot h_1 + \cdots + a_r \circ f \cdot h_r$$
$$\equiv a_0(0) + a_1(0) h_1 + \cdots + a_r(0) h_r,$$

for some functions $a_0, a_1, \ldots, a_r$, modulo $(f^*\mathfrak{m}_m)F^*\mathcal{E}_{m+r} + \mathfrak{m}_n^\infty \cap F^*\mathcal{E}_{m+r} \subseteq (f^*\mathfrak{m}_m)F^*\mathcal{E}_{m+r}$. Thus we see $1, h_1, \ldots, h_r$ generate $F^*\mathcal{E}_{m+r}/(f^*\mathfrak{m}_m)F^*\mathcal{E}_{m+r}$ over $\mathbf{R}$, and therefore they generate $\mathcal{R}_f$ over $\mathcal{E}_{\mathbf{R}^m,0}$ via f^*. $\qquad\square$

6. The cases of corank ≥ 2

If $\operatorname{corank}(f) \geq 2$, then the restriction of a versal opening of an unfolding of f is not necessarily a versal opening of f. That phenomenon was observed already in [15]. We utilise Proposition 5.2 if necessary to treat the following examples.

Example 6.1 (cf. Example 2.8). Let $f : (\mathbf{R}^2, 0) \to (\mathbf{R}^2, 0)$, $h(x, y) = (\frac{1}{2}x^2, \frac{1}{2}y^2) = (z, w)$. Then $\mathcal{R}_f$ is minimally generated by $1, x^3, y^3, x^3 y^3$ over $f^*\mathcal{E}_{\mathbf{R}^2,0}$. Therefore if we set $F : (\mathbf{R}^2, 0) \to (\mathbf{R}^5, 0)$ by

$$F(x, y) = (\frac{1}{2}x^2, \ \frac{1}{2}y^2, \ x^3, \ y^3, \ x^3 y^3),$$

then F is the mini-versal opening of f.

Here we give a concrete method to find the minimal generators as above. Let $h \in \mathcal{E}_{\mathbf{R}^2,0} = \mathcal{E}_2$. Then by the preparation theorem we have

$$h \equiv (a \circ f)x + (b \circ f)y + (c \circ f)xy, \ (\text{mod.} \ f^*\mathcal{E}_2).$$

The condition that $h \in \mathcal{R}_f$ is equivalent to that dh belongs to Jacobi module $\mathcal{J}_f$. We calculate

$$dh \equiv (a \circ f)dx + (b \circ f)dy + (c \circ f)(ydx + xdy), \ (\text{mod.} \ \mathcal{J}_f),$$

and set

$$(a \circ f)dx + (b \circ f)dy + (c \circ f)(ydx + xdy) = Axdx + Bydy,$$

for some function $A, B \in \mathcal{E}_2$. Again by the preparation theorem, we put

$$A = (a_1 \circ f) + (a_2 \circ f)x + (a_3 \circ f)y + (a_4 \circ f)xy,$$
$$B = (b_1 \circ f) + (b_2 \circ f)x + (b_3 \circ f)y + (b_4 \circ f)xy.$$

Then

$$Ax = (a_1 \circ f)x + (a_2 \circ f)x^2 + (a_3 \circ f)xy + (a_4 \circ f)x^2 y$$
$$= 2(za_2) + a_1 x + 2(za_4)y + a_3 \, xy,$$
$$By = (b_1 \circ f)y + (b_2 \circ f)xy + (b_3 \circ f)y^2 + (b_4 \circ f)xy^2$$
$$= 2(wb_3) + 2(wb_4) + b_1 y + b_2 xy.$$

omitting "$\circ f$", where $z = \frac{1}{2}x^2$ and $w = \frac{1}{2}y^2$. Then we have

$$a + cy = 2za_2 + a_1 x + (2za_4)y + a_3 xy,$$
$$b + cx = 2wb_3 + 2(wb_4) + b_1 y + b_2 xy$$

and therefore

$$(a - 2za_2) + (-a_1)x + (c - 2za_4)y + (-a_3)xy = 0,$$
$$(b - 2wb_3) + (c - 2wb_4)x + (-b_1)y + (-b_2)xy = 0.$$

Since $\mathcal{E}_2$ is free over $f^*\mathcal{E}_2$ in this example, we have

$$a = 2za_2, a_1 = 0, c = 2za_4, a_3 = 0, b = 2wb_3, c = 2wb_4, b_1 = 0, b_2 = 0.$$

Then we have $(a_4, b_4) = k(w, z)$ for a function $k \in \mathcal{E}_2$ and

$$h \equiv (2za_2 \circ f)x + (2wb_3 \circ f)y + (2zwk \circ f)xy, \ (\text{mod.} \ f^*\mathcal{E}_2).$$

Thus we find a minimal system of generators $1, 2zx, 2wy, 2zwxy$, namely $1, x^3, y^3, x^3 y^3$ of $\mathcal{R}_f$ over $f^*\mathcal{E}_2$.

In each case of the following three examples, we have the mini-versal openings using Proposition 5.2.

Example 6.2. Let $g : (\mathbf{R}^3, 0) \to (\mathbf{R}^3, 0)$ be a map-germ defined by

$$g(x, y, u) = (z, w, u) = (\frac{1}{2}x^2 + uy, \frac{1}{2}y^2 + ux, \ u),$$

which is an unfolding of f in Example 6.1. Then $\mathcal{R}_g$ is minimally generated over $g^*\mathcal{E}_{\mathbf{R}^3,0}$ by 1 and

$$\psi_3 = x^3 + y^3 + 3xyu,$$
$$\psi_5^{5,0} = x^5 + 5x^3 yu - 12x^2 u^3 + 9yu^4,$$
$$\psi_5^{0,5} = y^5 + 5xy^3 u - 12y^2 u^3 + 9xu^4,$$
$$\psi_6^{3,3} = x^3 y^3 - 12x^2 y^2 u^2 - 11x^3 u^3 - 11y^3 u^3 - 12xyu^4.$$

Therefore $i^*\mathcal{R}_g \subsetneq \mathcal{R}_f$, where $i : (\mathbf{R}^2, 0) \to (\mathbf{R}^3, 0), i(x, y) = (x, y, 0)$, and we see that g is not an extendable unfolding of f.

The versal opening of g is given by $G : (\mathbf{R}^3, 0) \to (\mathbf{R}^7, 0) = (\mathbf{R}^3 \times \mathbf{R}^4, 0)$,

$$G(x, y, u) = (g(x, y, u), \ x^3 + y^3 + 3xyu, \ x^5 + 5x^3 yu - 12x^2 u^3 + 9yu^4,$$
$$y^5 + 5xy^3 u - 12y^2 u^3 + 9xu^4, x^3 y^3 - 12x^2 y^2 u^2 - 11x^3 u^3 - 11y^3 u^3 - 12xyu^4).$$

Then

$$G(x, y, 0) = (\frac{1}{2}x^2, \frac{1}{2}y^2, x^3 + y^3, x^5, y^5, x^3 y^3)$$

is not a versal opening of $f = (\frac{1}{2}x^2, \frac{1}{2}y^2)$. Note that the element ψ_3 gives a Lagrange immersion of type D_4^+, which is a Lagrange stable lifting of g. Other elements are obtained by operating lowerable vector fields of g to ψ_3. See §7.

Example 6.3 (Hyperbolic case). Let $h : (\mathbf{R}^4, 0) \to (\mathbf{R}^4, 0)$ be the stable map-germ

$$h(x, y, \lambda, \mu) = (z, w, \lambda, \mu) = (\frac{1}{2}x^2 + y\lambda, \ \frac{1}{2}y^2 + x\mu, \ \lambda, \ \mu)$$

of $\mathcal{K}$-class $I_{2,2}$ ([22]). Then $\mathcal{R}_f$ is minimally generated over $h^*\mathcal{E}_{\mathbf{R}^4,0}$ by 1 and

$$\varphi_4 = x^3\mu + y^3\lambda + 3xy\lambda\mu$$

$$\varphi_5^{3,2} = x^3y^2 - 2x^2y\lambda\mu + x\lambda^2\mu^2$$

$$\varphi_5^{2,3} = x^2y^3 - 2xy^2\lambda\mu + y\lambda^2\mu^2$$

$$\varphi_5^{5,0} = x^5 + 5x^3y\lambda + 15y\lambda^3\mu$$

$$\varphi_5^{0,5} = y^5 + 5xy^3\mu + 15x\lambda\mu^3$$

$$\varphi_6 = x^3y^3 - 3xy\lambda^2\mu^2.$$

We have the mini-versal opening $H : (\mathbf{R}^4, 0) \to (\mathbf{R}^4 \times \mathbf{R}^6, 0) = (\mathbf{R}^{10}, 0)$ of h by

$$H = (f, \varphi_4, \varphi_5^{3,2}, \varphi_5^{2,3}, \varphi_5^{5,0}, \varphi_5^{0,5}, \varphi_6).$$

Moreover we see that

$$j^*\mathcal{R}_h \subsetneq \mathcal{R}_g(\subsetneq \mathcal{E}_{\mathbf{R}^3,0}), \qquad (j \circ i)^*\mathcal{R}_h \subsetneq i^*\mathcal{R}_g \subsetneq \mathcal{R}_f(\subsetneq \mathcal{E}_{\mathbf{R}^3,0}),$$

where $j : (\mathbf{R}^3, 0) \to (\mathbf{R}^4, 0), j(x, y, u) = (x, y, u, u)$. Thus the unfolding h of f is not extendable, which is also not extendable regarded as an unfolding of g as well.

Now we show the concrete way of calculations for Example 6.3 to make sure ourselves:

Let $k \in \mathcal{E}_{\mathbf{R}^4,0} = \mathcal{E}_4$. By the preparation theorem, we set,

$$k = (a_0 \circ h) + (a_1 \circ h)x + (a_2 \circ h)y + (a_3 \circ h)xy.$$

Then

$$dk \equiv ((a_1 \circ h) + (a_3 \circ h)y)dx + ((a_2 \circ h) + (a_3 \circ h)x)dy \pmod{\mathcal{J}_h}.$$

We suppose dk is equal to the form

$$Adz + Bdw + Cd\lambda + Dd\mu$$
$$= (Ax + B\mu)dx + (A\lambda + By)dy + (Ay + C)d\lambda + (Bx + D)d\mu.$$

Then we have

$$(a_1 \circ h) + (a_3 \circ h)y = Ax + B\mu, \quad (a_2 \circ h) + (a_3 \circ h)x = A\lambda + By,$$
$$Ay + C = 0, \quad Bx + D = 0.$$

Then $C = -Ay, D = -Bx$. Now set

$$A = (A_0 \circ h) + (A_1 \circ h)x + (A_2 \circ h)y + (A_3 \circ h)xy,$$
$$B = (B_0 \circ h) + (B_1 \circ h)x + (B_2 \circ h)y + (B_3 \circ h)xy.$$

Then we have

$$Ax = (2zA_1 - 4wA_3) + (A_0 + 4\lambda\mu A_3)x + (-2\lambda A_1 + 2zA_3)y + A_2xy,$$

omitting "$\circ h$". Similarly we have

$$By = (2wB_2 - 4z\mu B_3) + (-2\mu B_2 + 2wB_3)x + (B_0 + 4\lambda\mu B_3)y + B_1xy,$$
$$A\lambda = (\lambda A_0) + (\lambda A_1)x + (\lambda A_2)y + (\lambda A_3)xy.$$
$$B\mu = (\mu B_0) + (\mu B_1)x + (\mu B_2)y + (\mu B_3)xy.$$

Then we set, to find analytic or formal generators,

$$a_1 = 2zA_1 - 4w\lambda A_3 + \mu B_0, \quad a_3 = -2\lambda A_1 + 2zA_3 + \mu B_2,$$
$$0 = A_0 + 4\lambda\mu A_3 + \mu B_1, \quad 0 = A_2 + \mu B_3,$$

and

$$a_2 = 2wB_2 - 4z\mu B_3 + \lambda A_0, \quad a_3 = -2\mu B_2 + 2wB_3 + \lambda A_1,$$
$$0 = B_0 + 4\lambda\mu B_3 + \lambda A_2, \quad 0 = B_1 + \lambda A_3.$$

Then we are led to the relation:

$$3\lambda A_1 - 3\mu B_2 - 2zA_3 + 2wB_3 = 0 \cdots\cdots\cdots (*).$$

If A_1, A_2, B_2, B_3 satisfy the relation (*), then A_0, A_2, B_0, B_1 are determined from them and so $a_1 \circ h, a_2 \circ h, a_3 \circ h$:

$$A_0 = -3\lambda\mu A_3, \quad A_2 = -\mu B_3, \quad B_0 = -3\lambda\mu B_3, \quad B_1 = -\lambda A_3,$$

and

$$a_1 = 2zA_1 - 4w\lambda A_3 - 3\lambda\mu^2 B_3,$$
$$a_2 = -3\lambda^2\mu A_3 + 2wB_2 - 4z\mu B_3,$$
$$a_3 = -2\lambda A_1 + 2zA_3 + \mu B_2 (= \lambda A_1 - 2\mu B_2 + 2wB_3).$$

Since $(3\lambda, -3\mu, -2x, 2w)$ are regular sequence in $\mathcal{E}_4$, the first Koszul cohomology for them vanishes (see for instance, [23]). Then, by setting (A_1, B_2, A_3, B_3) as

$$(\mu, \lambda, 0, 0), \ (2z, \ 0, \ 3\lambda, \ 0), \ (2w, \ 0, \ 0, \ -3\lambda),$$
$$(0, \ 2z, \ -3\mu, \ 0), \ (0, \ 2w, \ 0, \ 3\mu), \ (0, \ 0, \ w, \ z),$$

respectively, we have elements $\varphi_4, \varphi_5^{3,2}, \varphi_5^{2,3}, \varphi_5^{5,0}, \varphi_5^{0,5}, \varphi_6$. such that

$$1, \varphi_4, \varphi_5^{3,2}, \varphi_5^{2,3}, \varphi_5^{5,0}, \varphi_5^{0,5}, \varphi_6$$

generate $\mathcal{R}_h^\omega$ over $h^*\mathcal{O}_4$. Then, by Proposition 5.2, they generate $\mathcal{R}_h$ over $h^*\mathcal{E}_4$.

Similarly we have the following.

Example 6.4 (Elliptic case). Let $k : (\mathbf{R}^4, 0) \to (\mathbf{R}^4, 0)$ be the stable map-germ given by

$$k(x, y, \lambda, \mu) = (\frac{1}{2}(x^2 - y^2) + \lambda x + \mu y, \ xy + \mu x - \lambda y, \ \lambda, \ \mu),$$

of $\mathcal{K}$-class $II_{2,2}$. Then $\mathcal{R}_k$ is minimally generated over $k^*\mathcal{E}_4$ by 1 and $\rho_4, \rho_5^{3,2}, \rho_5^{2,3}, \rho_5^{5,0}, \rho_5^{0,5}, \rho_6$, where

$$\rho = a_1 x + a_2 y + \tfrac{1}{2}a_3(x^2 + y^2),$$

$$a_1 = 2zA_1 + 2wA_2 + (-\tfrac{3}{2}\lambda^3 - \tfrac{3}{2}\lambda\mu^2 - 3z\lambda - 3w\mu)A_3$$
$$+(-\tfrac{3}{2}\lambda^2\mu - \tfrac{3}{2}\mu^3 - 3z\mu + 3w\lambda)B_3,$$

$$a_2 = -2wA_1 + 2zA_2 + (\tfrac{3}{2}\lambda^2\mu - \tfrac{3}{2}\mu^3 + z\mu - w\lambda)A_3$$
$$+(\tfrac{3}{2}\lambda^3 + \tfrac{3}{2}\lambda\mu^2 - z\lambda - 3w\mu)B_3,$$

$$a_3 = -\lambda A_1 - \mu A_2 + (z - \tfrac{1}{2}\lambda^2 + \tfrac{1}{2}\mu^2)A_3 + (w - \lambda\mu)B_3,$$

and $\rho = \rho_4, \rho_5^{3,2}, \rho_5^{2,3}, \rho_5^{5,0}, \rho_5^{0,5}, \rho_6$ respectively for

$$(A_1, A_2, A_3, B_3) = (\lambda, \mu, 0, 0),$$
$$(0, z - \tfrac{3}{2}\lambda^2 + \tfrac{3}{2}\mu^2, 0, 3\lambda),$$
$$(0, w - 3\lambda\mu, -3\lambda, 0),$$
$$(z - \tfrac{3}{2}\lambda^2 + \tfrac{3}{2}\mu^2, 0, 0, -3\mu),$$
$$(w - 3\lambda\mu, 0, 3\mu, 0),$$
$$(0, 0, z - \tfrac{3}{2}\lambda^2 + \tfrac{3}{2}\mu^2, w - 3\lambda\mu)).$$

Note that, in the process of calculations, we see that A_1, A_2, A_3, B_3 obey the relation

$$(3\mu)A_1 + (-3\lambda)A_2 + (-w + 3\lambda\mu)A_3 + (z - \frac{3}{2}\lambda^2 + \frac{3}{2}\mu^2)B_3 = 0.$$

and that $3\mu, -3\lambda, -w + 3\lambda\mu, z - \frac{3}{2}\lambda^2 + \frac{3}{2}\mu^2$ form a regular sequence in $\mathcal{E}_4$.

Theorem 6.5. *Any stable mono-germ* $(\mathbf{R}^4, a) \to (\mathbf{R}^4, b)$, *and therefore any stable multi-germ* $(\mathbf{R}^4, A) \to (\mathbf{R}^4, b)$ *has a versal opening.*

Proof. Let $f : (\mathbf{R}^4, a) \to (\mathbf{R}^4, b)$ be a stable map-germ. Then f is of corank ≤ 1 and is diffeomorphic to a Morin map or f is of corank 2 and is diffeomorphic to the germ h of Example 6.3 or k of Example 6.4 (see for instance [7]). In the case of corank one, we have constructed the versal opening in Proposition 4.1. In the case of corank two, we have constructed the versal opening, using a normal form of f in Examples 6.3 and Example 6.4. If a map-germ ℓ has a versal opening and f is diffeomorphic to ℓ, then f has a versal opening (cf. Lemma 2.2 (3)). Therefore f has a versal opening. Let $f : (\mathbf{R}^4, A) \to (\mathbf{R}^4, b)$ be a stable multi-germ. Then, for each $a_i \in A$, the restriction $f_i : (\mathbf{R}^4, a_i) \to (\mathbf{R}^4, b)$ of f to $(\mathbf{R}^n, a_i)$ is a stable germ. Since each f_i has a versal opening, we see that f itself has a versal opening by Remark 2.9. $\qquad\qquad\square$

7. Openings and lowerable vector fields

Let $f : (\mathbf{R}^n, A) \to (\mathbf{R}^m, b)$ a map-germ. A germ of vector field ξ over $(\mathbf{R}^n, A)$ is called *lowerable* for f, or f-*lowerable*, if there exists a germ of vector field η over $(\mathbf{R}^m, b)$ such that $f_*\xi = \eta \circ f$ as a germ of vector field along f. The lowerable vector fields form an $f^*\mathcal{E}_{\mathbf{R}^m, b}$-module, which is denoted by $\mathfrak{X}_f$.

Lemma 7.1 (cf. [14]). *Let* $f : (\mathbf{R}^n, A) \to (\mathbf{R}^m, b)$ *be a map-germ and* ξ *a lowerable vector field for* f. *Then* $\xi(f^*\mathcal{E}_{\mathbf{R}^m, b}) \subseteq f^*\mathcal{E}_{\mathbf{R}^m, b}$ *and* $\xi(\mathcal{R}_f) \subseteq \mathcal{R}_f$. *Therefore* $\mathfrak{X}_f(f^*\mathcal{E}_{\mathbf{R}^m, b}) \subseteq f^*\mathcal{E}_{\mathbf{R}^m, b}$ *and* $\mathfrak{X}_f(\mathcal{R}_f) \subseteq \mathcal{R}_f$.

Proof. Suppose $f_*\xi = \eta \circ f$ as in the above definition.

Let $a \in f^*\mathcal{E}_{\mathbf{R}^m, b}$. Then

$$\xi(f^*a)(x) = \langle d(f^*a)(x), \xi(x)\rangle = \langle f^*(da)(x), \xi(x)\rangle = \langle (da)(f(x)), f_*(\xi(x))\rangle$$
$$= \langle (da)(f(x)), \eta(f(x))\rangle = (\eta a)(f(x)) = (f^*(\eta a))(x).$$

Therefore $\xi(f^*a) = f^*(\eta a)$.

Let $b \in \mathcal{R}_f$. Then $db = \sum_{i=1}^m p_i df_i$ for some $p_i \in \mathcal{E}_{\mathbf{R}^n, A}$. Then we have

$$d(\xi b) = L_\xi(\sum_{i=1}^m p_i df_i) = \sum_{i=1}^m (\xi p_i) df_i + \sum_{i=1}^m p_i d(\xi f_i).$$

Since each $\xi f_i \in f^*\mathcal{E}_{\mathbf{R}^m, b} \subseteq \mathcal{R}_f$, we see $d(\xi b) \in \mathcal{J}_f$. Therefore $\xi b \in \mathcal{R}_f$. $\quad\square$

The differential operator $D : \mathcal{E}_{\mathbf{R}^n,A} \to \mathcal{E}_{\mathbf{R}^n,A}$ is called *lowerable* for $f : (\mathbf{R}^n, A) \to (\mathbf{R}^m, b)$, if D is a finite sum of operators of form

$$\xi_1 \xi_2 \cdots \xi_s$$

with coefficients in $f^* \mathcal{E}_{\mathbf{R}^n,b}$, where $\xi_1, \xi_2, \ldots, \xi_s$ are lowerable vector field (regarded as first order differential operators) for f. The lowerable differential operators form a (non-commutative) $f^* \mathcal{E}_{\mathbf{R}^n,b}$-algebra, which is denoted by $\mathfrak{L}_f$.

By Lemma 7.1, we have:

Corollary 7.2. *Let* $f : (\mathbf{R}^n, A) \to (\mathbf{R}^m, b)$ *be a map-germ and* $D : \mathcal{E}_{\mathbf{R}^n,A} \to \mathcal{E}_{\mathbf{R}^n,A}$ *be a lowerable differential operator for* f. *Then* $D(f^* \mathcal{E}_{\mathbf{R}^m,b}) \subseteq f^* \mathcal{E}_{\mathbf{R}^m,b}$ *and* $D(\mathcal{R}_f) \subseteq \mathcal{R}_f$. *Therefore* $\mathfrak{L}_f(f^* \mathcal{E}_{\mathbf{R}^m,b}) \subseteq f^* \mathcal{E}_{\mathbf{R}^m,b}$ *and* $\mathfrak{L}_f(\mathcal{R}_f) \subseteq \mathcal{R}_f$.

We conclude the present paper by examining how lowerable vector fields act on the ramification modules and the structure of mini-versal openings for Example 6.1, Example 6.2 and Example 6.3.

Example 7.3. The module of lowerable vector fields for f in Example 6.1 is generated by

$$\xi_1 = x \frac{\partial}{\partial x}, \quad \xi_2 = y \frac{\partial}{\partial y}$$

over $f^* \mathcal{E}_2$. Then we have

$$\xi_1(x^3) = 3x^3, \quad \xi_2(x^3) = 0,$$
$$\xi_1(y^3) = 0, \quad \xi_2(y^3) = 3y^3,$$
$$\xi_1(x^3 y^3) = 3x^3 y^3, \quad \xi_2(x^3 y^3) = 3x^3 y^3.$$

Define the module $S = \langle 1, x^3, y^3 \rangle_{f^* \mathcal{E}_m} \subsetneq \mathcal{R}_f$ over $f^* \mathcal{E}_m$. Then we see

$$\mathfrak{X}_f(S) = S, \quad \mathfrak{L}_f(S) = S.$$

The module of lowerable vector fields for g in Example 6.2 is generated by

$$\begin{cases} \xi_0 = x \frac{\partial}{\partial x} + y \frac{\partial}{\partial y} + u \frac{\partial}{\partial u}, \\ \xi_1 = (-\frac{3}{2} u^3 + \frac{1}{3} zx) \frac{\partial}{\partial x} + (-\frac{1}{2} u^2 x) \frac{\partial}{\partial y} + (-\frac{1}{3} zu) \frac{\partial}{\partial u}, \\ \xi_2 = (\frac{1}{3} wx + \frac{1}{2} u^2 y) \frac{\partial}{\partial x} + (\frac{3}{2} u^3 - \frac{1}{2} uxy) \frac{\partial}{\partial y} + \frac{2}{3} wu \frac{\partial}{\partial u}, \\ \xi_3 = (-\frac{3}{2} wu^2 - \frac{1}{2} zuy + \frac{1}{2} wxy) \frac{\partial}{\partial x} + (\frac{3}{2} u^3 + \frac{1}{2} zxy) \frac{\partial}{\partial y} + (-zw) \frac{\partial}{\partial u}, \end{cases}$$

over $g^* \mathcal{E}_2$, where $z = \frac{1}{2}x^2 + uy, w = \frac{1}{2}y^2 + ux$. Then we have that

$$1, \; \xi_0 \psi_3, \; \xi_1 \psi_3, \; \xi_2 \psi_3, \; \xi_3 \psi_3$$

generate $\mathcal{R}_g$ over $g^* \mathcal{E}_3$. This fact is a consequence of Lagrange stability of induced Lagrangian immersion from (g, ψ_3), $\tilde{g} : (\mathbf{R}^3, 0) \to T^* \mathbf{R}^3 = \mathbf{R}^6$ defined by

$$\tilde{g} = (z, w, u, p_1, p_2, p_3) = (z = \frac{1}{2}x^2 + uy, \; w = \frac{1}{2}y^2 + ux, \; u, \; 3x, \; 3y, \; -3xy).$$

See [11]. Therefore, if we consider the $g^* \mathcal{E}_3$-module T generated by $1, \psi_3$, then we have that

$$\mathfrak{X}_g(T) = \mathcal{R}_g, \quad \mathfrak{L}_g(T) = \mathcal{R}_g.$$

The lowerable vector fields for h in Example 6.3 is generated by

$$
\begin{cases}
\xi_0 = x\dfrac{\partial}{\partial x} + y\dfrac{\partial}{\partial y} + \lambda\dfrac{\partial}{\partial \lambda} + \mu\dfrac{\partial}{\partial \mu}, \\[2mm]
\xi_1 = x\dfrac{\partial}{\partial x} - y\dfrac{\partial}{\partial y} + (3\lambda)\dfrac{\partial}{\partial \lambda} + (-3\mu)\dfrac{\partial}{\partial \mu}, \\[2mm]
\xi_2 = (-3\lambda\mu + xy)\dfrac{\partial}{\partial x} + (-\mu x)\dfrac{\partial}{\partial y} + (-x^2 - 2\lambda y)\dfrac{\partial}{\partial \lambda}, \\[2mm]
\xi_3 = (-\lambda y)\dfrac{\partial}{\partial x} + (-3\lambda\mu + xy)\dfrac{\partial}{\partial y} + (-y^2 - 2\mu x)\dfrac{\partial}{\partial \mu}.
\end{cases}
$$

Then we have $\xi_0 \varphi_4 = 4\varphi_4, \xi_1 \varphi_4 = 0$ and

$$\xi_2 \varphi_4 = -\varphi_5^{2,3} - 8\lambda(w^2 + z\mu^2) \equiv -\varphi_5^{2,3} \; (\mathrm{mod.} \; h^* \mathcal{E}_4),$$
$$\xi_3 \varphi_4 = -\varphi_5^{3,2} - 8\mu(z^2 + w\lambda^2) \equiv -\varphi_5^{3,2} \; (\mathrm{mod.} \; h^* \mathcal{E}_4).$$

Moreover

$$\xi_0 \varphi_5^{3,2} = 5\varphi_5^{3,2}, \xi_0 \varphi_5^{2,3} = 5\varphi_5^{2,3}, \xi_0 \varphi_5^{5,0} = 5\varphi_5^{5,0}, \xi_0 \varphi_5^{0,5} = 5\varphi_5^{0,5},$$
$$\xi_1 \varphi_5^{3,2} = \varphi_5^{3,2}, \xi_1 \varphi_5^{2,3} = -\varphi_5^{2,3}, \xi_1 \varphi_5^{5,0} = 5\varphi_5^{5,0}, \xi_1 \varphi_5^{0,5} = -5\varphi_5^{0,5},$$

and

$$\xi_2\varphi_5^{3,2} = 3\varphi_6^{3,3} + 18\lambda\mu\varphi_4 - 36zw\lambda\mu - 3\lambda^3\mu^3$$
$$\equiv 3\varphi_6^{3,3} + 18\lambda\mu\varphi_4 \ (\text{mod. } h^*\mathcal{E}_4),$$

$$\xi_2\varphi_5^{2,3} = -4\lambda\varphi_5^{0,5} - 32z^2\mu^2 + 68w\lambda^2\mu^2$$
$$\equiv -4\lambda\varphi_5^{0,5} \ (\text{mod. } h^*\mathcal{E}_4),$$

$$\xi_2\varphi_5^{5,0} = 5\lambda\varphi_5^{3,2} - 16z^2\mu - 4w\lambda^2\mu$$
$$\equiv 5\lambda\varphi_5^{3,2} \ (\text{mod. } h^*\mathcal{E}_4),$$

$$\xi_2\varphi_5^{0,5} = 15\mu^2\varphi_4 - 60zw\mu^2 - 45\lambda^2\mu^4$$
$$\equiv 15\mu^2\varphi_4 \ (\text{mod. } h^*\mathcal{E}_4),$$

$$\xi_3\varphi_5^{3,2} = -4\mu\varphi_5^{5,0} - 32w^2\lambda^2 + 68z\lambda^3\mu^2$$
$$\equiv -4\mu\varphi_5^{5,0} \ (\text{mod. } h^*\mathcal{E}_4),$$

$$\xi_3\varphi_5^{2,3} = 3\varphi_6^{3,3} + 18\lambda\mu\varphi_4 - 36zw\lambda\mu - 3\lambda^3\mu^2$$
$$= 3\varphi_6^{3,3} + 18\lambda\mu\varphi_4 \ (\text{mod. } h^*\mathcal{E}_4),$$

$$\xi_3\varphi_5^{5,0} = 15\lambda^2\varphi_4 - 60zw\lambda^2 - 45\lambda^4\mu^2$$
$$\equiv 15\lambda^2\varphi_4 \ (\text{mod. } h^*\mathcal{E}_4),$$

$$\xi_3\varphi_5^{0,5} = 5\mu\varphi_5^{2,3} - 16w^2\lambda - 4z\lambda\mu^2$$
$$\equiv 5\mu\varphi_5^{2,3} \ (\text{mod. } h^*\mathcal{E}_4).$$

We consider the $h^*\mathcal{E}_4$-module U generated by $1, \varphi_4$ and V generated by $1, \varphi_4, \varphi_5^{3,2}, \varphi_5^{2,3}, \varphi_5^{5,0}, \varphi_5^{0,5}$. Then we have

$$\mathfrak{X}_h(U) = V, \quad \mathfrak{L}_h(U) = \mathcal{R}_h.$$

It would be an interesting open problem, for the geometry of openings, to find a submodule U, as small as possible, of $\mathcal{R}_f$ which satisfies $\mathfrak{X}_f(U) = \mathcal{R}_f$ or $\mathfrak{L}_f(U) = \mathcal{R}_f$, for any stable map-germ $f : (\mathbf{R}^n, A) \to (\mathbf{R}^m, b), n \geq 5$. In this paper we have solved the problem just for the case $n = 4, m = 4$. To understand the structure of openings in general case, it seems to be necessary to study more higher dimensional cases $n \geq 5$.

References

1. V.I. Arnol'd, *Catastrophe theory*, 3rd edition, Springer-Verlag, (1992).
2. E. Bierstone, P.D. Milman, *Relations among analytic functions I*, Ann. Inst. Fourier, Grenoble, **37-1** (1987), 187–239.

3. E. Bierstone, P.D. Milman, *Relations among analytic functions II*, Ann. Inst. Fourier, Grenoble, **37**–**2** (1987), 49–77.

4. T. Bröcker, *Differentiable Germs and Catastrophes*, London Math. Soc. Lecture Note Series **17**, Cambridge Univ. Press (1975).

5. J.W. Bruce, P.J. Giblin, *Curves and singularities, A geometrical introduction to singularity theory*, 2nd ed., Cambridge Univ. Press, (1992).

6. T. Gaffney, M.A. Vitulli, *Weak subintegral closure of ideals*, Adv. in Math., **226** (2011), 2089–2117.

7. M. Golubitsky, V. Guillemin, *Stable mappings and their singularities*, Graduate Texts in Math., Springer-Verlag (1974).

8. A.B. Givental', *Varieties of polynomials having a root of fixed co-multiplicity and generalized Newton equation*, Funct. Anal. Appl. **16**–**1** (1982), 13–18.

9. G. Ishikawa, *Families of functions dominated by singularities of mappings*, Master thesis, Kyoto University (1982), in Japanese. (A part of [9] is published as [10]).

10. G. Ishikawa, *Families of functions dominated by distributions of C-classes of mappings*, Ann. Inst. Fourier, **33**-**2** (1983), 199–217.

11. G. Ishikawa, *Lagrangian stability of a Lagrangian map-germ in a restricted class — open Whitney umbrellas and open swallowtails*, Proceedings of the Symposium "Singularity Theory and its Applications", 1989, Hokkaido Univ., ed. by G. Ishikawa, S. Izumiya and T. Suwa, Hokkaido Univ. Technical Report Series in Math., **12** (1989), 267–273.

12. G. Ishikawa, *Parametrization of a singular Lagrangian variety*, Trans. Amer. Math. Soc., **331**–**2** (1992), 787–798.

13. G. Ishikawa, *Parametrized Legendre and Lagrange varieties*, Kodai Math. J., **17**–**3** (1994), 442–451.

14. G. Ishikawa, *Developable of a curve and determinacy relative to osculation-type*, Quart. J. Math. Oxford, **46** (1995), 437–451.

15. G. Ishikawa, *Symplectic and Lagrange stabilities of open Whitney umbrellas*, Invent. math., **126** (1996), 215–234.

16. G. Ishikawa, *Singularities of tangent varieties to curves and surfaces*, Journal of Singularities, **6** (2012), 54–83.

17. G. Ishikawa, *Tangent varieties and openings of map-germs*, to appear in RIMS Kōkyūroku Bessatsu.

18. G. Ishikawa, S. Janeczko, *Symplectic bifurcations of plane curves and isotropic liftings*, Quarterly J. Math. Oxford, **54** (2003), 73–102.

19. T. de Jong, D. van Straten, *A deformation theory for nonisolated singularities*, Anh. Math. Sem. Univ. Hamburg, **60** (1990), 177–208.

20. B. Malgrange, *Ideals of Differentiable Functions*, Oxford Univ. Press (1966).

21. J. Martinet, *Singularities of Smooth Functions and Maps*, London Math. Soc. Lecture Note Series **58**, Cambridge Univ. Press (1982).

22. J.N. Mather, *Stability of C^∞ mappings IV: The nice dimensions*, Lecture Notes in Math. **192**, Springer (1971), pp.192–253.

23. H. Matsumura, *Commutative Algebra*, Benjamin (1970).

24. D. Mond, *Deformations which preserve the non-immersive locus of a map-germ*, Math. Scand., **66** (1990), 21–32.

25. R. Pellikaan, *Finite determinacy of functions with non-isolated singularities,* Proc. London Math . Soc., **57** (1988), 357–382.

26. C.T.C. Wall, *Finite determinacy of smooth map-germs,* Bull. London Math. Soc. **13** (1981), 481–539.

Acknowledgement: This paper has the origin in my master thesis [9] [10]. From the time when I was an undergraduate student of Kyoto University under the supervision of Professor Masahisa Adachi, Professor Satoshi Koike kept giving me kind encouragements and friendship throughout. I would like to thank him in the occasion of his 60th birthday.

I am grateful to the referee for helpful comments to correct and improve the original manuscript.

Non concentration of curvature near singular points of two variable analytic functions

Satoshi Koike, Tzee-Char Kuo and Laurentiu Paunescu

Department of Mathematics, Hyogo University of Teacher's Education,
Hyogo, Japan
koike@hyogo-u.ac.jp

School of Mathematics, University of Sydney, Sydney, NSW, 2006, Australia
tck@maths.usyd.edu.au

School of Mathematics, University of Sydney, Sydney, NSW, 2006, Australia
laurent@maths.usyd.edu.au

In this paper we study the phenomenon of non concentration of curvature of level curves of two variable complex analytic function germs, and we characterise it in terms of tree models and topological types. In addition, we also discuss the relationship in the real case, between the phenomenon of non concentration of curvature and real tree models (or blow-analytic types). In particular, we give an example to demonstrate that the corresponding characterisation does not hold in the real case.

Keywords: concentration of curvature, tree model, blow-analyticity, curvature tableland.
AMS classification numbers: 14B05, 14H50, 32S15, 58K20

1. Introduction

Let $f : (\mathbb{K}^2, 0) \to (\mathbb{K}, 0)$ be an analytic function germ not identically zero, where $\mathbb{K} = \mathbb{R}$ or $\mathbb{C}$. The singular point set of f is contained in the zero-set $f^{-1}(0)$. What kind of relationship can we find between the singularity type of $f : \mathbb{K}^2 \to \mathbb{K}$ and its level curves $f = c$, $0 < |c| < \epsilon, \epsilon$ small? In this respect N. A'Campo made a profound observation. R. Langevin explored the A'Campo phenomenon and found (in [16]) an interesting relationship between the integration of the total Gaussian curvatures of the level curves of a complex analytic function and its Milnor number. In addition, E. Garcia Barroso and B. Teissier analysed the concentration of curvature of the level curves of a complex analytic function in [1], and J.-J. Risler investigated the curvature problem for the real Milnor fibre in [18]. In a couple of previous papers ([9], [10]), using the language of infinitesimals as introduced in [14], [15], we studied the A'Campo's curvature bumps for both real and complex analytic functions.

The curvature formula in the real case is

$$K_f^{\mathbb{R}}(x,y) := \pm \frac{\Delta_f(x,y)}{(f_x(x,y)^2 + f_y(x,y)^2)^{\frac{3}{2}}},$$

and that in the complex case is

$$K_f^{\mathbb{C}}(z,w) := -\frac{2|\Delta_f(z,w)|^2}{(|f_z(z,w)|^2 + |f_w(z,w)|^2)^3},$$

where

$$\Delta_f := 2f_x f_y f_{xy} - f_x^2 f_{yy} - f_y^2 f_{xx}.$$

These are known formulae for computing the Gaussian curvature of level curves $f = c$, $0 < |c| < \epsilon$ (cf. John A. Thorpe [19]).

Let us consider the polynomial functions f_1, $f_2 : (\mathbb{R}^2, 0) \to (\mathbb{R}, 0)$ defined by

$$f_1(x,y) = x^2 - y^3, \quad f_2(x,y) = x^4 - y^5.$$

Using the real formula above, we can easily see that if $|c|$ is sufficiently small, the level curves of f_1 and f_2 are very close to be vertical in a (wide) horn-neighbourhood of the x-axis. Similarly the level curves of f_2 are also nearly horizontal in a horn-neighbourhood of the y-axis. Therefore the union of the level curves $f_2 = \pm c$, $c \neq 0$, looks like a rectangle outside some thin horn-like region tangent to the y-axis. Intuitively the concentration of curvature of f_2 happens in this thin region. On the other hand, we can see that the concentration of curvature of f_1 happens in a thin horn-neighbourhood of the y-axis. These phenomena are illustrated in the pictures below. (See also [9].)

In [6] the first named author and A. Parusiński gave a complete blow-analytic classification of two variable real analytic function germs in terms of their real tree model. See §3 for the definitions of blow-analytic equivalence and real tree model. The real tree models of the above f_1 and f_2 are drawn as follows:

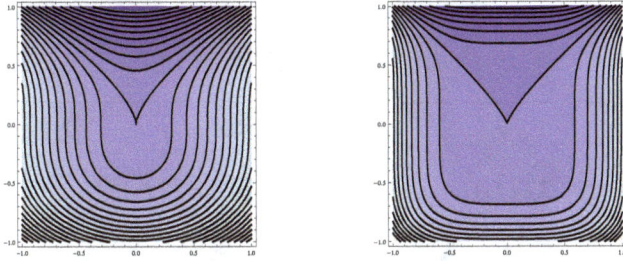

Fig. 1. $f_1 = c$, $f_2 = c$

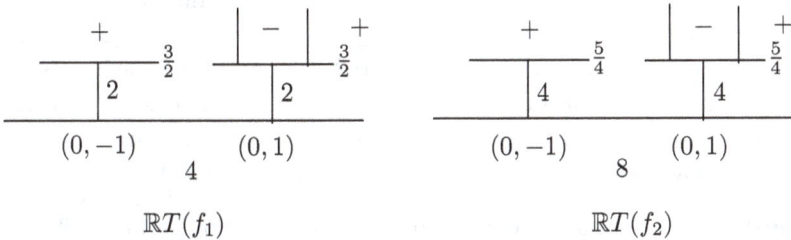

$\mathbb{R}T(f_1)$

$\mathbb{R}T(f_2)$

By an easy computation of the curvature using the above formula, we can observe that the concentration of curvature of f_1 and f_2 happens on the bars of height $\frac{3}{2}$ and $\frac{5}{4}$, respectively. Note that, in both cases, the concentration of curvature does not happen directly on the ground bar.

Remark 1.1. In [9] we made a more detailed analysis of the curvature of the level curves of the above f_1 and f_2. We let the trunks supporting the bars of height $\frac{3}{2}$ in $\mathbb{R}T(f_1)$ and of height $\frac{5}{4}$ in $\mathbb{R}T(f_2)$ grow upward and put provisional bars of height 2 and $\frac{4}{3}$ on the grown trunks in $\mathbb{R}T(f_1)$ and $\mathbb{R}T(f_2)$, respectively. In each case the concentration of curvature happens on the provisional bar. We do not elaborate on this observation in this paper.

In the real case the zero set $f^{-1}(0)$ can be just $\{(0,0)\}$ as a set germ at the origin. Let us consider such polynomial functions f_3, $f_4 : (\mathbb{R}^2, 0) \to (\mathbb{R}, 0)$ defined by

$$f_3(x, y) = x^4 + y^4, \quad f_4(x, y) = x^4 + y^6.$$

The real tree models of f_3 and f_4 are drawn below.

$$\begin{array}{cc}
\underbrace{+}_{8} & \underbrace{\overset{\displaystyle -\frac{3}{2}}{\boxed{4}}\ \overset{\displaystyle +}{}\ \overset{\displaystyle \frac{3}{2}}{\boxed{4}}}_{8} \\
(0,-1) \qquad (0,1) \\
\mathbb{R}T(f_3) & \mathbb{R}T(f_4)
\end{array}$$

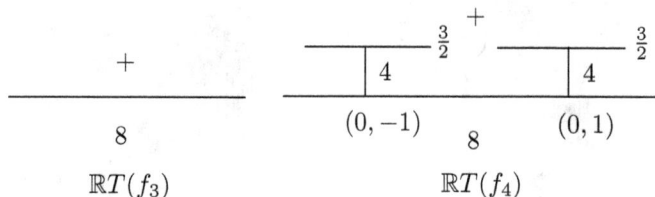

In the case of f_3 we can see that $|K_{f_3}(x,y)|$ on each level curve takes a maximum along some curves near $y = \pm x$, but non concentration of curvature happens. Note that there is no bar of height bigger than 1 in $\mathbb{R}T(f_3)$. On the other hand, a level curve $f_4 = c$, $c > 0$, looks like a rectangle outside some thin horn-like region tangent to the y-axis. The concentration of curvature of f_4 happens in this thin region. We can see that the concentration of curvature of f_4 happens on the bars of height $\frac{3}{2}$.

The above observations give rise to natural questions.

Question 1.2. For a two variable real analytic function germ $f : (\mathbb{R}^2, 0) \to (\mathbb{R}, 0)$ is there any relationship between its real tree model and the concentration of curvature?

Question 1.3. For a two variable complex analytic function germ $f : (\mathbb{C}^2, 0) \to (\mathbb{C}, 0)$ is there any relationship between its tree model and the concentration of curvature?

In this paper we give an affirmative answer to Question 1.3. Namely, we show that non concentration of curvature happens if and only if no bar of height bigger than 1 appears in the tree model $T(f)$ (Theorem 5.4). This condition is also equivalent to the homogeneous-likeness. (See §3 for the definition of homogeneous-like.) From this result, we can see that the appearance of concentration of curvature is a topological invariant in the complex case (Corollary 5.5).

On the other hand, we have a negative answer to Question 1.2. More precisely, the corresponding condition on the real tree model, which is equivalent to the homogeneous-likeness, implies non concentration of curvature, but the converse is not valid. Indeed, we give an example to demonstrate that non concentration of curvature does not always imply homogeneous-likeness (Proposition 5.7).

In order to introduce our notions of concentration of curvature and non concentration of curvature, we need to also consider the curvature of the level curve $f = 0$ in a punctured neighbourhood of $0 \in \mathbb{K}^2$, even if the zero-set is a singular locus of f. The sign of the curvature is not essential for these notions. In the next section we define the non-directed curvature for the level curves $f = c$, $|c| < \epsilon$, in a punctured neighbourhood of $0 \in \mathbb{K}^2$, and using this definition, we introduce the notions of concentration of curvature and non concentration of curvature in §5. In §3 we recall the notions of tree model and real tree model, mention some related results, and give a characterisation of the homogeneous-likeness in terms of tree models. In §4 we review several results on A'Campo curvature bumps proved in [9,10], which are necessary for the proofs of our main result in the complex case mentioned in §5.

Throughout our paper we will use the following convention and notations. We call an analytic function germs $f : (\mathbb{K}^2, 0) \to (\mathbb{K}, 0)$ *mini-regular* in x, if

$$f(x,y) := H_m(x,y) + H_{m+1}(x,y) + \cdots, \quad H_m(1,0) \neq 0,$$

where $m = O(f)$ is the order of f and $H_k(x,y)$ are homogeneous forms of degree $k \geq m$.

For two non-negative functions f, $g : [0, \delta) \to [0, \infty)$, $\delta > 0$,

(i) we write $f \approx g$ if there exist $0 < K_1 \leq K_2$ and $0 < \delta_1 \leq \delta$ such that

$$K_1 g(\epsilon) \leq f(\epsilon) \leq K_2 g(\epsilon)$$

for $0 \leq \epsilon \leq \delta_1$, and

(ii) we write $f \ll g$ if $g(\epsilon) > 0$ for $0 < \epsilon < \delta_1$ with $\delta_1 \leq \delta$ and the quotient $\frac{f(\epsilon)}{g(\epsilon)}$ tends to 0 as $\epsilon \to 0$.

2. Non-directed curvature

In this section we introduce the non-directed curvature for level curves of an analytic function germ, including the zero locus. We first consider the real case.

Let $g : (\mathbb{R}^2, 0) \to (\mathbb{R}, 0)$ be an irreducible analytic function germ, and let $f = g^m$ for $m \geq 1$. Then we have

$$\Delta_f = m^3 g^{3(m-1)} \Delta_g \quad \text{and} \quad (f_x^2 + f_y^2)^{\frac{3}{2}} = \pm m^3 g^{3(m-1)} (g_x^2 + g_y^2)^{\frac{3}{2}}.$$

Therefore, after cancellation of $|g^{3(m-1)}|$, we have

$$|K_f| = \frac{|\Delta_f|}{(f_x^2 + f_y^2)^{\frac{3}{2}}} = \frac{|\Delta_g|}{(g_x^2 + g_y^2)^{\frac{3}{2}}} = |K_g|$$

in a punctured neighbourhood of $0 \in \mathbb{R}^2$. Note that in case m is odd, $K_f = K_g$ after cancellation of $g^{3(m-1)}$.

Let $f(x, y) = x^2 y$ and $g(x, y) = xy$, then $|K_f|$ does not coincide with $|K_g|$ in a punctured neighbourhood of $0 \in \mathbb{R}^2$. Nevertheless we still have some reasonable results on comparing their curvature. We first prepare some lemmas.

Lemma 2.1. Let $g : (\mathbb{R}^2, 0) \to (\mathbb{R}, 0)$ be an irreducible analytic function such that $g^{-1}(0) \neq \{0\}$ as germs at $0 \in \mathbb{R}^2$. Then $g_x^2 + g_y^2$ is not divisible by g.

Proof. Since g is irreducible, g has an isolated singularity at $0 \in \mathbb{R}^2$ (cf. [11]). Suppose that $g_x^2 + g_y^2$ is divisible by g. Then $g_x = 0$ and $g_y = 0$ along $g^{-1}(0)$. This contradicts the isolated singularity of g. $\qquad \square$

Remark 2.2. We cannot drop the assumption that $g^{-1}(0) \neq \{0\}$ at $0 \in \mathbb{R}^2$. Let $g(x, y) = x^2 + y^2$, g is irreducible and $g^{-1}(0) = \{0\}$. On the other hand, $g_x^2 + g_y^2 = 4(x^2 + y^2)$ is divisible by g.

Let $f, g, h : (\mathbb{R}^2, 0) \to (\mathbb{R}, 0)$ be analytic function germs such that $f = g^m h$ for $m \geq 2$. Suppose that g is an irreducible analytic function such that $g^{-1}(0) \neq \{0\}$ as germs at $0 \in \mathbb{R}^2$ and that h is not divisible by g. Then we have the following lemmas.

Lemma 2.3. $|\Delta_f|$ and $(f_x^2 + f_y^2)^{\frac{3}{2}}$ are divisible by $|g^{3(m-1)}|$.

Proof. By an easy computation, we have

$$f_x = (mg_x h + gh_x)g^{m-1},$$
$$f_y = (mg_y h + gh_y)g^{m-1},$$
$$f_{xx} = (mg_{xx}gh + m(m-1)g_x^2 h + 2mg_x gh_x + g^2 h_{xx})g^{m-2},$$
$$f_{xy} = (mg_{xy}gh + m(m-1)g_x g_y h + mg_x gh_y + mg_y gh_x + g^2 h_{xy})g^{m-2},$$
$$f_{yy} = (mg_{yy}gh + m(m-1)g_y^2 h + 2mg_y gh_y + g^2 h_{yy})g^{m-2}.$$

Then we have

$$\Delta_f = \{m^3 h^3 (2g_x g_y g_{xy} - g_x^2 g_{yy} - g_y^2 g_{xx}) + Hg\}g^{3(m-1)},$$
$$f_x^2 + f_y^2 = \{m^2 h^2 (g_x^2 + g_y^2) + 2mh(g_x h_x + g_y h_y)g + (h_x^2 + h_y^2)g^2\}g^{2(m-1)},$$

where $H : \mathbb{R}^2 \to \mathbb{R}$ is an analytic function germ at $0 \in \mathbb{R}^2$. Therefore $|\Delta_f|$ and $(f_x^2 + f_y^2)^{\frac{3}{2}}$ are divisible by $|g^{3(m-1)}|$. $\qquad \square$

After this, we put $\widetilde{\Delta}_f := \dfrac{|\Delta_f|}{|g^{3(m-1)}|}$ and $f_\nabla := \dfrac{(f_x^2 + f_y^2)^{\frac{3}{2}}}{|g^{3(m-1)}|}$.

Lemma 2.4. (1) $f_\nabla \neq 0$ *over* $g^{-1}(0) \smallsetminus \{0\}$.

(2) *Let* $\tilde{K}_f := \frac{\tilde{\Delta}_f}{f_\nabla}$. *Then* $\tilde{K}_f = |K_g|$ *over* $g^{-1}(0) \smallsetminus \{0\}$.

Proof. (1) By Lemmas 2.1 and 2.3, $m^2 h^2 (g_x^2 + g_y^2) \neq 0$ over $g^{-1}(0) \smallsetminus \{0\}$. Therefore $f_\nabla \neq 0$ over $g^{-1}(0) \smallsetminus \{0\}$.

(2) Over $g^{-1}(0) \smallsetminus \{0\}$,

$$\tilde{K}_f = \frac{|m^3 h^3 (2g_x g_y g_{xy} - g_x^2 g_{yy} - g_y^2 g_{xx})|}{(m^2 h^2 (g_x^2 + g_y^2))^{\frac{3}{2}}}$$

$$= \frac{|2g_x g_y g_{xy} - g_x^2 g_{yy} - g_y^2 g_{xx}|}{(g_x^2 + g_y^2)^{\frac{3}{2}}} = |K_g|. \qquad \square$$

Remark 2.5. Lemma 2.4 holds also for $m = 1$. In particular, (2) becomes the following:

$K_f = \pm K_g$ over $g^{-1}(0) \smallsetminus \{0\}$, where the sign $\pm$ depends on the sign of h at $(x, y) \in g^{-1}(0) \smallsetminus \{0\}$.

Let $f : (\mathbb{R}^2, 0) \to (\mathbb{R}, 0)$ be an analytic function germ. Then f has a decomposition of the following form:

$$f = f_1^{m_1} \cdots f_k^{m_k} h, \quad m_i \geq 1 \ (1 \leq i \leq k), \ k \in \mathbb{N} \cup \{0\}, \ f_i \neq f_j \ (i \neq j), \quad (2.1)$$

where each $f_i : (\mathbb{R}^2, 0) \to (\mathbb{R}, 0)$, $1 \leq i \leq k$, is an irreducible analytic component of f such that $f_i^{-1}(0) \neq \{0\}$ as germs at $0 \in \mathbb{R}^2$, and $h : (\mathbb{R}^2, 0) \to (\mathbb{R}, 0)$ is an analytic function germ such that $h^{-1}(0) \subseteq \{0\}$ as germs at $0 \in \mathbb{R}^2$. Note that $k \in \mathbb{N}$ in case h is a unit.

Let us define the *non-directed curvature* of level curves of f as follows:

$$\overline{K}_f^{\mathbb{R}}(x, y) := \begin{cases} |K_f^{\mathbb{R}}(x, y)| & \text{if } (x, y) \in \mathbb{R}^2 \smallsetminus f^{-1}(0) \\ |K_{f_j}(x, y)| & \text{if } (x, y) \in f_j^{-1}(0) \smallsetminus \{0\} \ (1 \leq j \leq k). \end{cases}$$

The next proposition follows from Lemmas 2.3 and 2.4.

Proposition 2.6. $\overline{K}_f^{\mathbb{R}}$ *is continuous in a punctured neighbourhood of* $0 \in \mathbb{R}^2$.

As mentioned in the introduction, using the concept of non-directed curvature, we shall define the notions of concentration of curvature and non concentration of curvature for two variable real analytic function germs in §5.

Concerning the reduction of the problem of concentration of curvature, it may be natural to ask the following question.

Question 2.7. Let $f : (\mathbb{R}^2, 0) \to (\mathbb{R}, 0)$ be an analytic function germ with a decomposition of form (2.1), and let $g : (\mathbb{R}^2, 0) \to (\mathbb{R}, 0)$ be the function germ defined by $g = f_1 \cdots f_k h$. One can ask whether there are positive numbers $0 < C_1 < C_2$ such that

$$C_1 |K_f| \leq |K_g| \leq C_2 |K_f| .$$

If the answer would be affirmative, then it would be enough to consider only reduced analytic function germs. Unfortunately this does not hold in general. In fact, we have

Example 2.1. Let f, $g : (\mathbb{R}^2, 0) \to (\mathbb{R}, 0)$ be real analytic function germs defined by

$$f(x, y) = x^2(x - y^2), \quad g(x, y) = x(x - y^2).$$

By simple computations, we can see

$$|K_f| = \frac{|\Delta_f|}{(f_x^2 + f_y^2)^{\frac{3}{2}}} \approx \frac{1}{|y|^3}, \quad |K_g| = \frac{|\Delta_g|}{(g_x^2 + g_y^2)^{\frac{3}{2}}} \approx 1$$

on the curve $\{x = \frac{2}{3}y^2\}$. Therefore there does not exist $C_1 > 0$ such that $C_1 |K_f| \leq |K_g|$.

We next consider the complex case.

Lemma 2.8. Let $g : (\mathbb{C}^2, 0) \to (\mathbb{C}, 0)$ be an irreducible analytic function germ. Then g has an isolated singularity at $0 \in \mathbb{C}^2$.

Let f, g, $h : (\mathbb{C}^2, 0) \to (\mathbb{C}, 0)$ be analytic function germs such that $f = g^m h$ for $m \geq 2$. Suppose that g is irreducible and h is not divisible by g. Then we have the following.

Lemma 2.9. $|\Delta_f|^2$ and $(|f_z|^2 + |f_w|^2)^3$ are divisible by $|g|^{6(m-1)}$.

Proof. Similarly to Lemma 2.3, we have

$$\Delta_f = \{m^3 h^3 (2g_x g_y g_{xy} - g_x^2 g_{yy} - g_y^2 g_{xx}) + Hg\} g^{3(m-1)},$$
$$|f_z|^2 + |f_w|^2 = (|mg_z h + gh_z|^2 + |mg_w h + gh_w|^2)|g|^{2(m-1)},$$

where $H : \mathbb{C}^2 \to \mathbb{C}$ is an analytic function germ at $0 \in \mathbb{C}^2$. It follows that $|\Delta_f|^2$ and $(|f_z|^2 + |f_w|^2)^3$ are divisible by $|g|^{6(m-1)}$. $\qquad \square$

After this, we put $\widetilde{\Delta}_f^{\mathbb{C}} := -\frac{2|\Delta_f|^2}{|g|^{6(m-1)}}$ and $f_\triangledown^{\mathbb{C}} := \frac{(|f_z|^2 + |f_w|^2)^3}{|g|^{6(m-1)}}$. Using the same arguments as in Lemma 2.4, we can show a similar lemma.

Lemma 2.10. (1) $f_{\nabla}^{\mathbb{C}} \neq 0$ over $g^{-1}(0) \setminus \{0\}$.

(2) Let $\widetilde{K}_f^{\mathbb{C}} := \frac{\widetilde{\Delta}_f^{\mathbb{C}}}{f_{\nabla}^{\mathbb{C}}}$. Then $\widetilde{K}_f^{\mathbb{C}} = K_g^{\mathbb{C}}$ over $g^{-1}(0) \setminus \{0\}$.

Let $f : (\mathbb{C}^2, 0) \to (\mathbb{C}, 0)$ be an analytic function germ. Then f has a decomposition of the following form:

$$f = f_1^{m_1} \cdots f_k^{m_k}, \quad m_i \geq 1 \ (1 \leq i \leq k), \ k \in \mathbb{N}, \ f_i \neq f_j \ (i \neq j),$$

where each $f_i : (\mathbb{C}^2, 0) \to (\mathbb{C}, 0)$, $1 \leq i \leq k$, is an irreducible analytic component of f.

Let us define the curvature of level curves of f as follows:

$$\widehat{K}_f^{\mathbb{C}}(z, w) := \begin{cases} K_f^{\mathbb{C}}(z, w) & \text{if } (z, w) \in \mathbb{C}^2 \setminus f^{-1}(0) \\ K_{f_j}^{\mathbb{C}}(z, w) & \text{if } (z, w) \in f_j^{-1}(0) \setminus \{0\} \ \ (1 \leq j \leq k). \end{cases}$$

The next proposition follows from Lemmas 2.9 and 2.10.

Proposition 2.11. $\widehat{K}_f^{\mathbb{C}}$ is continuous in a punctured neighbourhood of $0 \in \mathbb{C}^2$.

When we discuss the phenomenon of concentration of curvature and non concentration of curvature for a complex analytic function germ $f : (\mathbb{C}^2, 0) \to (\mathbb{C}, 0)$, the minus sign is not essential. Accordingly we set

$$\overline{K}_f^{\mathbb{C}}(z, w) := |\widehat{K}_f^{\mathbb{C}}(z, w)|$$

over a punctured neighbourhood of $0 \in \mathbb{C}^2$. In §5 we shall use this definition of $\overline{K}_f^{\mathbb{C}}$ to define the notions of concentration or non concentration of curvatures in the complex case.

3. Tree model and real tree model

In this section we briefly review the definitions of tree model and real tree model introduced in [12] and [6,7], respectively.

We first recall the notion of tree model. This is a kind of geometric interpretation of the classical Zariski theorem on the topology of complex plane curves ([20]).

Let $f(x, y)$ be a complex analytic function germ of multiplicity m and mini-regular in x. Let $x = \lambda_i(y)$, $i = 1, \ldots, m$, be the complex Newton-Puiseux roots of f. We denote the contact order at zero of λ_i and λ_j by

$$O(\lambda_i, \lambda_j) := \mathrm{ord}_0 (\lambda_i - \lambda_j)(y).$$

Let $h \in \mathbb{Q}$. We say that λ_i, λ_j are congruent modulo h^+ if $O(\lambda_i, \lambda_j) > h$.

The *tree model* $T(f)$ of f is defined as follows. Draw a vertical line segment as the *main trunk* of the tree. Mark $m = \text{mult}_0 f(x, y)$ alongside the trunk. Let $h_0 := \min\{O(\lambda_i, \lambda_j) \mid 1 \leq i, j \leq m\}$. Then draw a bar, B_0, on top of the main trunk. Call $h(B_0) := h_0$ the *height* of B_0. The roots are divided into equivalence classes modulo h_0^+. We then represent each equivalence class by a vertical line segment drawn on top of B_0, and call it *trunk*. If a trunk consists of s roots we say it has *multiplicity s*, and mark s alongside. The same construction is repeated recursively. The construction terminates at the stage where all trunks contain only single, possibly multiple, roots of f.

Example 3.1. Let $f : (\mathbb{C}^2, 0) \to (\mathbb{C}, 0)$ be a polynomial function defined by

$$f(x, y) = x(x + y)^2 (x^2 - y^3)(x^3 + y^7)^3.$$

The tree model $T(f)$ is drawn as follows:

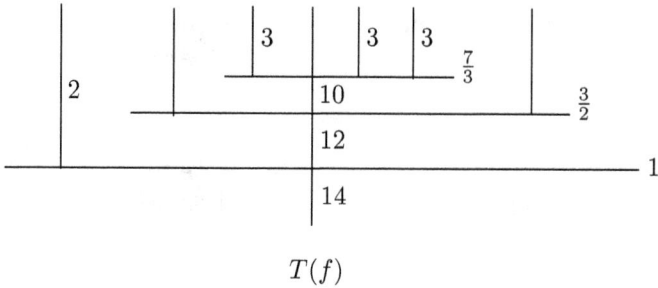

$$T(f)$$

We do not take care on the position of trunks on bars in the complex case.

We call the sets of roots corresponding to trunks *bunches*. Each bunch A is a set of roots growing through a unique bar $B(A)$. Fix a bunch A with finite height $h(A)$. Take a root $\lambda_i(y) \in A$. Let $\lambda_A(y)$ denote $\lambda_i(y)$ with all terms y^e, $e \geq h(A)$, omitted. Then we can write $\lambda_i(y) \in A$ as

$$\lambda_i(y) = \lambda_A(y) + c_i y^{h(A)} + \cdots, \quad c_i \in \mathbb{C}. \tag{3.2}$$

We next recall the notion of real tree model. Let $f(x, y)$ be a real analytic function germ. Consider the Newton-Puiseux roots as arcs $x = \lambda_i(y)$ defined for $y \in \mathbb{R}$, $y \geq 0$. The complex conjugation acts on the Newton-Puiseux roots, and hence on the tree model $T(f)$. A bunch A of $T(f)$ is called *real* if it is stable under complex conjugation. A bar or a trunk is *real* if and only if so are the corresponding bunches growing on it. We denote by $T_+(f)$ the conjugation invariant part of $T(f)$.

Fix v a unit vector of $\mathbb{R}^2$. Fix any local system of coordinates x, y such that $f(x, y)$ is mini-regular in x and v is of the form (v_1, v_2) with $v_2 > 0$. Consider the Newton-Puiseux roots of f as arcs $x = \lambda_i(y)$ defined for $y \in \mathbb{R}$, $y \geq 0$.

We define *the real tree model of f relative to v*, denoted by $\mathbb{R}T_v(f)$, as the part of $T_+(f)$ consisting only of those roots tangent to v with the following additional information. Let A be a real bunch such that $B = B(A)$ is a bar of $\mathbb{R}T_v(f)$. Then:

- draw the trunks on B realising the sub-bunches of A keeping the clockwise order of the roots (i.e. the order of the coefficients c_i in (3.2)),
- whenever B gives a new Puiseux pair of some roots of A we mark 0 on B and draw from it the unique sub-bunch of A with $c_i = 0$, i.e. consisting of the roots that do not have the new Puiseux pair at B. Hence we are also able to determine from the tree the sub-bunches with positive and negative c_i. Graphically, we identify $0 \in B$ with the point of B that belongs to the trunk supporting B.

The *real tree model $\mathbb{R}T(f)$ of f* is defined as follows:

- Draw a bar B_0 (identified with S^1). We define $h(B_0) = 1$ and call B_0 *the ground bar*. We mark $m(B_0) := 2\,\mathrm{mult}_0\, f(x, y)$ below the ground bar. We call $m(B_0)$ the *multiplicity* of the ground bar.
- Draw on B_0 the non-trivial $\mathbb{R}T_v(f)$ for $v \in S^1$, keeping the clockwise order.
- Let v_1, v_2 be any two subsequent unit vectors for which $\mathbb{R}T_v(f)$ is non-trivial. Mark on B_0 of $\mathbb{R}T(f)$ the sign of f in the sector between v_1 and v_2.

Definition 3.1. We call two real tree models $\mathbb{R}T(f)$, $\mathbb{R}T(g)$ *isomorphic*, if there is a homeomorphism φ between the trees which maps the ground bar to the ground bar preserving the multiplicity, bars to bars preserving the heights, trunks to trunks preserving the multiplicities, and the signs of the characteristic coefficients, after we move trunks, if necessary, on the bars whose heights are not giving Puiseux pairs.

Blow-analytic equivalence is a notion defined by the second author in [13] as a natural equisingularity condition for real analytic function germs : $(\mathbb{R}^n, 0) \to (\mathbb{R}, 0)$. We say that two real analytic function germs $f : (\mathbb{R}^2, 0) \to$

$(\mathbb{R}, 0)$ and $g : (\mathbb{R}^2, 0) \to (\mathbb{R}, 0)$ are *blow-analytically equivalent,* if there exist compositions of finite point blowings-up $\mu : (M, \mu^{-1}(0)) \to (\mathbb{R}^2, 0)$, $\mu' : (M', \mu'^{-1}(0)) \to (\mathbb{R}^2, 0)$ and an analytic isomorphism $\Phi : (M, \mu^{-1}(0)) \to (M', \mu'^{-1}(0))$ which induces a homeomorphism $\phi : (\mathbb{R}^2, 0) \to (\mathbb{R}^2, 0)$ such that $f = g \circ \phi$.

Remark 3.2. In general blow-analytic equivalence is defined using *real modifications* ([13]). In the two variable case a real modification is attained by a composition of finite point blowings-up ([6]). For properties on blow-analyticity, see the surveys [3], [5].

Blow-analytic equivalence of two variable real analytic function germs is completely determined by the real tree models as follows:

Theorem 3.3 ([6]). *Let $f : (\mathbb{R}^2, 0) \to (\mathbb{R}, 0)$ and $g : (\mathbb{R}^2, 0) \to (\mathbb{R}, 0)$ be real analytic function germs. Then f and g are blow-analytically equivalent if and only if the real tree models of f and g are isomorphic.*

Example 3.2. Let $f, g : (\mathbb{R}^2, 0) \to (\mathbb{R}, 0)$ be two polynomial functions defined by

$$f(x, y) = x^3 - y^4, \quad g(x, y) = x^3 + y^4.$$

The real tree models $\mathbb{R}T(f)$ and $\mathbb{R}T(g)$ are drawn as follows:

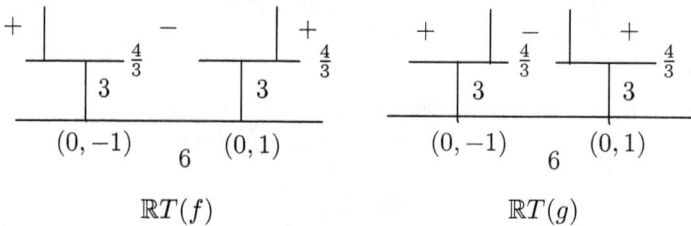

$$\mathbb{R}T(f) \qquad\qquad \mathbb{R}T(g)$$

Since $\mathbb{R}T(f)$ and $\mathbb{R}T(g)$ are not isomorphic, we can see by Theorem 3.3 that f and g are not blow-analytically equivalent.

Definition 3.4. We call a real analytic function germ $f : (\mathbb{R}^2, 0) \to (\mathbb{R}, 0)$ *homogeneous-like,* if f is blow-analytically equivalent to a homogeneous polynomial function germ.

Concerning the homogeneous-likeness of two variable real analytic function germs, we give a characterisation in terms of real tree models.

Lemma 3.5. *For a real analytic function germ* $f : (\mathbb{R}^2, 0) \to (\mathbb{R}, 0)$, *the following are equivalent.*

(1) f *is homogeneous-like.*

(2) Up to isomorphisms of real tree models, no bar except the ground bar appears in the real tree model $\mathbb{R}T(f)$.

Proof. Suppose that f is homogeneous-like, namely f is blow-analytically equivalent to a two variable real homogeneous polynomial function germ, and by Theorem 3.3, their real tree models are isomorphic. Since any two variable real homogeneous polynomial function germ clearly has a real tree model consisting of only the ground bar with finite trunks (including also the empty case) on it, (2) follows immediately.

On the other hand, consider a real analytic function germ $f : (\mathbb{R}^2, 0) \to (\mathbb{R}, 0)$ having a real tree model $\mathbb{R}T(f)$ which consists of only the ground bar with finite trunks on it. Taking into account also the multiplicities of the ground bar and the trunks and signs, it is easy to construct a two variable real homogeneous polynomial function having the real tree model isomorphic to $\mathbb{R}T(f)$. Therefore (1) follows immediately from Theorem 3.3.
□

We say that two complex analytic function germs $f : (\mathbb{C}^2, 0) \to (\mathbb{C}, 0)$ and $g : (\mathbb{C}^2, 0) \to (\mathbb{C}, 0)$ are *topologically equivalent*, if there exists a homeomorphism $\phi : (\mathbb{C}^2, 0) \to (\mathbb{C}^2, 0)$ such that $f = g \circ \phi$.

Now we recall the improved version of the Zariski theorem on the topology of two variable complex analytic function germs.

Theorem 3.6 (O. Zariski [20], Kuo - Lu [12], A. Parusiński [17]).
Let $f, g : (\mathbb{C}^2, 0) \to (\mathbb{C}, 0)$ *be analytic function germs. Then* f *and* g *are topologically equivalent if and only if the tree models of* f *and* g *coincide.*

Definition 3.7. We call a complex analytic function germ $f : (\mathbb{C}^2, 0) \to (\mathbb{C}, 0)$ *homogeneous-like*, if f is topologically equivalent to a homogeneous polynomial function germ.

Using Theorem 3.6 and a similar argument to Lemma 3.5, we can also characterise the homogeneous-likeness of a two variable complex analytic function germ as follows:

Lemma 3.8. *For a complex analytic function germ* $f : (\mathbb{C}^2, 0) \to (\mathbb{C}, 0)$, *the following are equivalent.*

(1) f *is homogeneous-like.*

(2) No bar of height bigger than 1 appears in the tree model $T(f)$.

Remark 3.9. Any real tree model has at least the ground bar. Some tree models in the complex case, however, can consist of only one trunk without any bar. This case also satisfies condition (2) in the above lemma.

4. Computation of A'Campo bumps

In [9,10] we introduced and computed A'Campo bumps of two variable analytic function germs using infinitesimals. Here we review some notions and results which will be used to show our main results.

Take an analytic function germ

$$\alpha : (\mathbb{C},0) \longrightarrow (\mathbb{C}^2,0), \quad \alpha(t) \not\equiv 0.$$

Let $\alpha_* := Im(\alpha)$ be the image set germ. Being an irreducible curve germ in $\mathbb{C}^2$, it has a unique tangent $T(\alpha_*)$ at 0, $T(\alpha_*)$ is a point of the Riemann Sphere $\mathbb{C}P^1$. We call α_* an *infinitesimal* at $T(\alpha_*)$. The *Enriched Riemann Sphere* is $\mathbb{C}P_*^1 := \{\alpha_*\}$.

The absolute value of the curvature computed *along* α_*, if not zero, can be written as

$$|K_f^{\mathbb{C}}(\alpha(t))| = as^L + \cdots, \quad a > 0, \quad L \in \mathbb{Q},$$

where $s = s(t)$ is the arc length. This is dominated by the leading term as^L as $s \to 0$.

If $|K_f^{\mathbb{C}}| \equiv 0$ along α_*, we write $(a,L) := (0,\infty)$. Hence we introduce the notations

$$(a,L) := a\delta^L, \quad 0_\nu := 0\delta^\infty, \quad \mathcal{V}(\mathbb{R}) := \{a\delta^L \mid a \neq 0\} \cup \{0_\nu\},$$

where δ is a symbol.

A *lexicographic ordering* on $\mathcal{V}(\mathbb{R})$ is defined: 0_ν is the smallest element, and

$$a\delta^L > a'\delta^{L'} \text{ if and only if either } L < L', \text{ or else } L = L', a > a'.$$

The *curvature function* K_* on $\mathbb{C}P_*^1$, and the component L_* are defined as follows:

$$K_* : \mathbb{C}P_*^1 \to \mathcal{V}(\mathbb{R}), \quad \alpha_* \to a\delta^L; \quad L_* : \mathbb{C}P_*^1 \to \mathbb{Q}, \quad \alpha_* \to L.$$

Recall the field $\mathbb{F}$ of convergent fractional power series in an indeterminate y is algebraically closed. A non-zero element of $\mathbb{F}$ is a convergent series

$$\alpha(y) = a_0 y^{n_0/N} + \cdots + a_i y^{n_i/N} + \cdots, \quad n_0 < n_1 < \cdots \,; \; n_i \in \mathbb{Z},$$

where $0 \neq a_i \in \mathbb{C}$, $N \in \mathbb{Z}^+$, $GCD(N, n_0, n_1, ...) = 1$. The *conjugates* of α are

$$\alpha^{(k)}_{conj}(y) := \sum a_i \theta^{kn_i} y^{n_i/N}, \quad 0 \leq k \leq N-1, \quad \theta := e^{\frac{2\pi\sqrt{-1}}{N}}.$$

The *order* is $O_y(\alpha) := n_0/N$, $O_y(0) := +\infty$. The *Puiseux multiplicity* is $m_{puis}(\alpha) := N$. The following are integral domains:

$$\mathbb{D}_1 := \{\alpha \in \mathbb{F} \mid O_y(\alpha) \geq 1\}, \quad \mathbb{D}_{1+} := \{\alpha \mid O_y(\alpha) > 1\},$$

having quotient field $\mathbb{F}$.

As in Projective Geometry, $\mathbb{C}P^1_*$ is a union of two charts: $\mathbb{C}P^1_* = \mathbb{C}_* \cup \mathbb{C}'_*$,

$$\mathbb{C}_* := \{\beta_* \in \mathbb{C}P^1_* \mid T(\beta_*) \neq [1:0]\}, \quad \mathbb{C}'_* := \{\beta_* \in \mathbb{C}P^1_* \mid T(\beta_*) \neq [0:1]\}.$$

Next we define the *contact order* $\mathcal{C}_{ord}(\alpha_*, \beta_*)$. We can assume $\alpha_*, \beta_* \in \mathbb{C}_*$. Then

$$\mathcal{C}_{ord}(\alpha_*, \beta_*) := \begin{cases} \infty & \text{if } \alpha_* = \beta_*, \\ \max_{i,j}\{O_y(\alpha^{(i)}_{conj}(y) - \beta^{(j)}_{conj}(y))\} & \text{if } \alpha_* \neq \beta_*. \end{cases}$$

The *horn subspaces* of $\mathbb{C}P^1_*$ centered at α_* of *order* e, e^+ are

$$\mathcal{H}_e(\alpha_*) := \{\beta_* \mid \mathcal{C}_{ord}(\alpha_*, \beta_*) \geq e\}, \quad \mathcal{H}_{e+}(\alpha_*) := \{\beta_* \mid \mathcal{C}_{ord}(\alpha_*, \beta_*) > e\},$$

respectively. In particular, $\mathcal{C}_{ord}(\alpha_*, \beta_*) = 1$ if $T(\alpha_*) \neq T(\beta_*)$; and $\mathcal{H}_1(\alpha_*) = \mathbb{C}P^1_*$ for all α_*. When there is no need to specify α_*, we write $\mathcal{H}_e := \mathcal{H}_e(\alpha_*)$.

Let $\mathcal{H}_e(\alpha_*)$ be given. If $\eta(y) = \alpha(y) + [cy^e + \cdots]$, c a *generic* number, then $L_*(\eta_*)$ is a constant. We write this constant as $L_*(\mathcal{H}^{grc}_e(\alpha_*))$. A *horn interval* of *radius* r, $r > 0$, is, by definition,

$$\mathcal{H}_e(\alpha_*, r) := \{\beta_* \mid \beta(y) = \alpha(y) + (cy^e + \cdots), \ |c| \leq r\}.$$

Definition 4.1. A horn subspace $\mathcal{H}_e(\alpha_*)$ is a *curvature tableland* if

(1) $\beta_* \in \mathcal{H}_e(\alpha_*) \implies L_*(\beta_*) \geq L_*(\mathcal{H}^{grc}_e(\alpha_*))$; and
(2) in the case $e > 1$, there exists e', $1 \leq e' < e$, such that

$$\nu_* \in \mathcal{H}_{e'}(\alpha_*) \setminus \mathcal{H}_e(\alpha_*) \implies L_*(\nu_*) > L_*(\mathcal{H}^{grc}_e(\alpha_*)).$$

Example 4.1. Let $f(x, y) = x^2 - y^5$. Then we have

$$f_x = 2x, \quad f_y = -5y^4, \quad f_{xx} = 2, \quad f_{xy} = 0, \quad f_{yy} = -20y^3.$$

Therefore we have

$$K^{\mathbb{C}}_f(x, y) = \frac{200|y|^6|8x^3 - 5y^5|^2}{(4|x|^2 + 25|y|^8)^3}.$$

We consider the horn subspace $\mathcal{H}_4(0_*)$. Then we can see

$$L_*(\mathcal{H}_4^{grc}(0_*)) = -8 = L_*(\beta_*) \text{ for all } \beta_* \in \mathcal{H}_4(0_*).$$

Let $0 < \epsilon < \frac{3}{2}$. If $\nu_* \in \mathcal{H}_{4-\epsilon}(0_*) \smallsetminus \mathcal{H}_4(0_*)$, then

$$-8 < L_*(\nu_*) \leq -8 + 6\epsilon.$$

It follows that $L_*(\nu_*) > L_*(\mathcal{H}_4^{grc}(0_*))$. Thus $\mathcal{H}_4(0_*)$ is a curvature tableland.

Definition 4.2. Let $\mathcal{H}_e$ be a curvature tableland. Take $\beta_* \in \mathcal{H}_e$. We say K_* has an *A'Campo bump* on $\mathcal{H}_{e+}(\beta_*)$, or simply say $\mathcal{H}_{e+}(\beta_*)$ is an *A'Campo bump*, if there exists $\epsilon > 0$ such that

$$\mu_* \in \mathcal{H}_e(\beta_*, \epsilon) \implies K_*(\beta_*) \geq K_*(\mu_*).$$

Let us apply a generic unitary transformation so that f is *mini-regular* in z where $m = O(f)$. Then the initial form H_m is written as follows:

$$H_m(z, w) = c(z - c_1 w)^{m_1} \cdots (z - c_r w)^{m_r}; \ m_i \geq 1; \ c_i \neq c_j; \text{ if; } i \neq j, \ (4.3)$$

where $1 \leq r \leq m$, $\sum m_i = m$, $c \neq 0$. Thus $H_m(z, w)$ is *degenerate* if and only if $r < m$.

Let ζ_i denote the Newton-Puiseux roots of $f(z, w)$, and γ_j those of f_z:

$$f(z, w) = unit \cdot \prod_{i=1}^{m}(z - \zeta_i(w)), \quad f_z(z, w) = unit \cdot \prod_{j=1}^{m-1}(z - \gamma_j(w)),$$

where $O_w(\zeta_i)$, $O_w(\gamma_j) \geq 1$. Each γ_j is called a *polar*, and so is γ_{j*}.

Definition 4.3. Given a polar γ, let $d_{gr}(\gamma)$ denote the *smallest* number e such that

$$O_w(\|Grad \ f(\gamma(w), w)\|) = O_w(\|Grad \ f(\gamma(w) + uw^e, w)\|),$$

where $u \in \mathbb{C}$ is a *generic* number. We call $d_{gr}(\gamma)$ the *gradient order* of γ. The $\mathbb{D}$-*gradient canyon* of γ, and the $*$-*gradient canyon* of γ_* are

$$\mathcal{G}(\gamma) := \{\alpha \in \mathbb{D}_1 \mid O_y(\alpha - \gamma) \geq d\}, \quad \mathcal{G}_*(\gamma_*) := \mathcal{H}_d(\gamma_*), \quad d := d_{gr}(\gamma),$$

respectively. When there is no confusion, we call them "canyons"; we also write

$$d := d_{gr}(\gamma), \quad \mathcal{G} := \mathcal{G}(\gamma), \quad \mathcal{G}_* := \mathcal{G}_*(\gamma_*).$$

The *degree* and *multiplicity* of $\mathcal{G}$, $\mathcal{G}_*$ are, respectively,

$$d_{gr}(\mathcal{G})v := d_{gr}(\mathcal{G}_*) := d,$$

$$m(\mathcal{G}) := \sharp\{k \mid \mathcal{G}(\gamma_k) = \mathcal{G}\},$$
$$m(\mathcal{G}_*) := \sharp\{k \mid \mathcal{G}_*(\gamma_{k*}) = \mathcal{G}_*\}.$$

Finally, we say $\mathcal{G}(\gamma)$ and $\mathcal{G}_*(\gamma_*)$ are *minimal* if

$$\mathcal{G}(\gamma_j) \subseteq \mathcal{G}(\gamma) \implies \mathcal{G}(\gamma_j) = \mathcal{G}(\gamma).$$

Let γ be a polar with $d < \infty$. We now define $L_\gamma \in \mathbb{Q}$, and a rational function $R_\gamma(u)$, $u \in \mathbb{C}$. We can assume $\gamma \in \mathbb{D}_{1+}$ so that $T(\gamma_*) = [0:1]$. If $d > 1$, define L_γ, $R_\gamma(u)$ by

$$|K_f^{\mathbb{C}}(\gamma(y) + uy^d, y)| := 2R_\gamma(u)y^{2L_\gamma} + \cdots, \quad R_\gamma(u) \not\equiv 0,$$

where y can be considered as the arc length of $(\gamma(y)+uy^d)_*$, since $\lim y/s = 1$.

In the case $d = 1$, define L_γ and $R_\gamma(u)$ by

$$\left|K_f^{\mathbb{C}}\left(\gamma(y) + \frac{uy}{\sqrt{1+|u|^2}}, \frac{y}{\sqrt{1+|u|^2}}\right)\right| := 2R_\gamma(u)y^{2L_\gamma} + \cdots, \quad R_\gamma(u) \not\equiv 0.$$

Theorem 4.4 ([9]). *A minimal $*$-canyon with $d < \infty$ is a curvature table-land, and vice versa.*

Take a minimal $\mathcal{G}(\gamma)$, $d < \infty$. Take a local maximum $R_\gamma(c)$ of $R_\gamma(u)$. Then $\mathcal{H}_{d+}(\gamma_*^{+c})$ is an A'Campo bump, where $\gamma^{+c}(y) := \gamma(y) + cy^d$.

All A'Campo bumps can be found in this way.

Addendum 4.5. ([9]) Every $\mathcal{G}(\gamma)$ with $d > 1$ is minimal. In this case,

$$m(\mathcal{G}) = \sharp\{k \mid \gamma_k \in \mathcal{G}\}, \quad m(\mathcal{G}_*) = \sharp\{k \mid \gamma_{k*} \in \mathcal{G}_*\}.$$

Let r be as in (4.3). There are exactly $r - 1$ polars of gradient degree 1; moreover,

$$d_{gr}(\gamma) = 1 \implies \mathcal{G}(\gamma) = \mathbb{D}_1, \; ; m(\mathcal{G}(\gamma)) = r - 1.$$

A minimal $\mathcal{G}(\gamma)$ with $d = 1$ exists if and only if $H_m(z, w)$ is non-degenerate. In this case every polar has $d = 1$, $\mathbb{C}P_*^1$ is the only curvature tableland and the only $*$-canyon.

Next we recall the Newton polygon relative to a polar. Let γ be a given polar, not a multiple root of $f(z, w)$, i.e., $f(\gamma(w), w) \neq 0$. We can apply a unitary transformation, if necessary, so that $T(\gamma_*) = [0:1]$, $\gamma \in \mathbb{D}_{1+}$.

Let us change coordinates:

$$Z := z - \gamma(w), \quad W := w, \quad F(Z, W) := f(Z + \gamma(W), W).$$

Then

$$\Delta_f(z,w) = \Delta_F(Z,W) + \gamma''(W)F_Z^3, \quad \|Grad_{z,w}f\| \approx \|Grad_{Z,W}F\|.$$

Recall that a monomial term aZ^iW^q, $a \neq 0$, $q \in \mathbb{Q}$, is represented by a "Newton dot" at (i,q). We shall simply say (i,q) is a *dot*.

If $i \geq 1$, then (i,q) is a dot of $F(Z,W)$ if and only if $(i-1,q)$ is one of F_Z. Since γ is a polar, F_Z has no dot of the form $(0,q)$; $F(Z,W)$ has no dot of the form $(1,q)$.

As $f(\gamma(w),w) \neq 0$, we know $F(0,W) \neq 0$. Hence

$$F(0,W) := aW^h + \cdots, \quad a \neq 0, \quad h = O_W(F(0,W)),$$

and then $(0,h)$ is a vertex of the Newton polygon $\mathcal{NP}(F)$, $(0,h-1)$ is one of the Newton polygon $\mathcal{NP}(F_W)$.

Let E_{top} denote the top edge of the Newton polygon $\mathcal{NP}(F)$, *i.e.*, the edge with left vertex $(0,h)$. Let (m_{top},q_{top}) denote the right vertex of E_{top}, and θ_{top} the angle of E_{top},

$$\tan\theta_{top} = \text{co-slope of } E_{top},$$

where the *co-slope* of a line passing through $(x,0)$ and $(0,y)$ is, by definition, y/x.

Let $(m'_{top},q'_{top}) \neq (0,h)$ be the dot of F on E_{top} which is *closest* to $(0,h)$. Then, clearly,

$$2 \leq m'_{top} \leq m_{top}, \quad \frac{h-q'_{top}}{m'_{top}} = \frac{h-q_{top}}{m_{top}} = \tan\theta_{top}.$$

Lemma 4.6 ([9]). *Let $\mathcal{L}^*$ denote the line joining $(0,h-1)$ (which is not a dot of F_Z) and a dot of F_Z such that no dot of F_Z lies below $\mathcal{L}^*$. Let σ^* denote the co-slope of $\mathcal{L}^*$. Then*

$$\tan d_{gr}(\gamma) = \sigma^* \geq \tan\theta_{top},$$

where $\sigma^* = \tan\theta_{top}$ if and only if $\sigma^* = 1$.

In order to show the main result in the complex case (Theorem 5.4), we shall use the above Theorem 4.4, Addendum 4.5 and Lemma 4.6.

5. Characterisations of no concentration of curvature

Let $\mathbb{K} = \mathbb{R}$ or $\mathbb{C}$. Consider the families of convergent Puiseux demi-arcs on $\mathbb{R}^2$ and the families of convergent Puiseux arcs on $\mathbb{C}^2$

$$\alpha_a : x = a_0 y + a_1 y^{\frac{n_1}{N}} + \cdots + a_i y^{\frac{n_i}{N}} + \cdots,$$

$$\text{or} \quad \alpha_{(0,b)} : y = b_1 x^{\frac{n_1}{N}} + \cdots + b_i x^{\frac{n_i}{N}} + \cdots,$$

where $1 < \frac{n_1}{N} < \frac{n_2}{N} \cdots$, $a_i, b_i \in \mathbb{K}$, $n_i, N \in \mathbb{Z}^+$. Here the demi-arc means that they are defined for $y \in \mathbb{R}$, $y \geq 0$ or $x \in \mathbb{R}$, $x \geq 0$. Let us denote this family by $\mathcal{A}$.

Let $f : (\mathbb{K}^2, 0) \to (\mathbb{K}, 0)$ be an analytic function germ not identically zero, mini-regular in x. For $\gamma \in \mathcal{A}$, we define the *curvature exponent along* γ by

$$e(\gamma) := O(\overline{K}_f^{\mathbb{K}}(\gamma)).$$

Then we define the *curvature exponent of* f by

$$\widehat{e}(f) := \inf_{\gamma \in \mathcal{A}} \{e(\gamma)\}.$$

Remark 5.1. We note that

$$-\widehat{e}(f) = \min\{s \mid |(x,y)|^s |\overline{K}_f^{\mathbb{K}}| \lesssim 1\}.$$

Therefore $\widehat{e}(f)$ is attained by some arc in $\mathcal{A}$.

Now we define the notion of concentration of curvature. Let $S := \{v \in \mathbb{K}^2 \mid \|v\| = 1\}$.

Definition 5.2. (1) We say that f has *concentration of curvature* at $0 \in \mathbb{K}^2$, if the set

$$\{v \in S \mid \exists \gamma \in \mathcal{A} \text{ s.t. } \lim_{t \to 0} \frac{\gamma(t)}{\|\gamma(t)\|} = v \ \& \ e(\gamma) = \widehat{e}(f)\}$$

is finite.

(2) We say that f has *no concentration of curvature* at $0 \in \mathbb{K}^2$, if f does not have concentration of curvature at $0 \in \mathbb{K}^2$.

Lemma 5.3. *Let $f : (\mathbb{K}^2, 0) \to (\mathbb{K}, 0)$ be a homogeneous-like analytic function germ. Then f has no concentration of curvature.*

Proof. We show only the real case. The complex case follows similarly.

Since f is homogeneous-like, it follows from Theorem 3.3 that there is a homogeneous polynomial $H : (\mathbb{R}^2, 0) \to (\mathbb{R}, 0)$ such that $\mathbb{R}T(f)$ is isomorphic to $\mathbb{R}T(H)$. Therefore f has a decomposition of the following form:

$$f = f_1^{m_1} \cdots f_k^{m_k} h, \ m_i \geq 0 \ (1 \leq i \leq k), \ k \in \mathbb{N} \cup \{0\},$$

where $f_i(x, y) = x - c_i y + \phi_i(x, y)$ with $j^1 \phi_i(0, 0) = 0$, and $c_1 < c_2 < \cdots < c_k$, and either $In(h)^{-1}(0) = \{0\}$ as germs at $0 \in \mathbb{R}^2$ or h is a unit. Here $In(h)$ means the initial homogeneous form of h.

Let $\widetilde{\Delta}_f$ be the numerator of $\widetilde{K}_f$ after cancellation of

$$f_1^{3(m_1-1)} \cdots f_k^{3(m_k-1)}$$

as in subsection 2. If $\widetilde{\Delta}_f \equiv 0$, then non concentration of curvature happens.

We next consider the directions $(e, 1)$, $e \in \mathbb{R}$, and $(1, 0)$. Note that if $\widetilde{\Delta}_f$ is not identically 0, then there are only finitely many e's such that $In(\widetilde{\Delta})(e, 1) = 0$.

- In the case where $In(\widetilde{\Delta}_f)(e, 1) \neq 0$, $\overline{K}_f^{\mathbb{R}} \approx \frac{1}{|y|}$ along any convergent Puiseux arc α_a with $a_0 = e$.
- In the case where $In(\widetilde{\Delta}_f)(e, 1) = 0$, $\overline{K}_f^{\mathbb{R}} \ll \frac{1}{|y|}$ along any convergent Puiseux arc α_a with $a_0 = e$.

Similarly for convergent Puiseux arcs with direction $(1, 0)$, we can see the following:

- In the case where $In(\widetilde{\Delta}_f)(1, 0) \neq 0$, $\overline{K}_f^{\mathbb{R}} \approx \frac{1}{|x|}$ for any $\alpha_{(0,b)}$.
- In the case where $In(\widetilde{\Delta}_f)(1, 0) = 0$, $\overline{K}_f^{\mathbb{R}} \ll \frac{1}{|x|}$ for any $\alpha_{(0,b)}$.

Therefore non concentration of curvature happens. □

Using the above lemmas, we give some characterisations of the non concentration of curvature in the complex case.

Theorem 5.4. *For an analytic function germ $f : (\mathbb{C}^2, 0) \to (\mathbb{C}, 0)$ not identically zero, the following are equivalent.*

(1) f is homogeneous-like.

(2) No bar of height bigger than 1 appears in the tree model $T(f)$.

(3) f has no concentration of curvature at $0 \in \mathbb{C}^2$.

Proof. By Lemmas 3.8 and 5.3, it suffices to show (3) implies (1). Taking a unitary transformation if necessary, we may assume that f is mini-regular in z.

Suppose that there exists a bar of height bigger than 1 in the tree model $T(f)$. Then it follows from the Kuo-Lu theorem ([12]) that there is a polar γ which leaves a trunk on this bar. Consider the Newton polygon relative to that polar. The top edge E_{top} has co-slope bigger than 1, because the co-slope is the height of the bar. Hence, by Lemma 4.6, the gradient order d of γ is bigger than 1. By Addendum 4.5 and Theorem 4.4, $\mathcal{G}(r)$ with $d > 1$ is minimal and $\mathcal{G}(\gamma)$ is a curvature tableland. If f has another polar γ' with gradient order $d' = 1$, $\mathcal{G}(\gamma')$ with $d' = 1$ is not minimal by Addendum 4.5. It follows from Theorem 4.4 that $\mathcal{G}(\gamma')$ is not a curvature tableland. Thus f has concentration of curvature at $0 \in \mathbb{C}^2$. □

The curvature of a complex analytic function itself can change under a non-unitary linear transformation. Nevertheless we can note the following.

Corollary 5.5. *For the two variable complex analytic function germs, the appearance of concentration of curvature is a topological invariant.*

By Lemmas 3.5 and 5.3 we have the following result in the real case.

Theorem 5.6. *For an analytic function germ $f : (\mathbb{R}^2, 0) \to (\mathbb{R}, 0)$ not identically zero, the following are equivalent.*

(1) f is homogeneous-like.

(2) Up to isomorphisms of real tree models, no bar except the ground bar appears in the real tree model $RT(f)$.

In addition, the above equivalent conditions imply

(3) f has no concentration of curvature at $0 \in \mathbb{R}^2$.

In the real case non concentration of curvature does not always imply the homogeneous-likeness. We give an example to demonstrate it. In order to see it, we prepare some notations. Let g, $h : [0, \delta) \to \mathbb{R}$ be convergent fractional power series functions, where h is not identically zero. Then we write $g \sim h$ if $\frac{g(y)}{h(y)}$ tends to 1 as $y \to +0$. If $g(y) - h(y)$ consists of terms with higher orders than the order of $h(y)$, we write $g(y) = h(y) + HOT$. Note that this HOT is not the high order terms in the usual sense.

Proposition 5.7. *Let $f : (\mathbb{R}^2, 0) \to (\mathbb{R}, 0)$ be a polynomial function defined by*

$$f(x, y) = (x - y^2)^3 + x^2 y^6 - xy^8 + y^{12}.$$

Then the non-directed curvature $\overline{K}_f^{\mathbb{R}}$ tends 2 along any convergent Puiseux demi-arc α_a and $\alpha_{(0,b)}$ on $\mathbb{R}^2$ as $y \to +0$ and $x \to +0$ respectively, namely non concentration of curvature happens with a constant coefficient. On the other hand, f is not homogeneous-like.

Proof. Let us express α_a and $\alpha_{(0,b)}$ as follows:

$$\alpha_a : \ x = a_0 y^{s_0} + a_1 y^{s_1} + a_2 y^{s_2} + \cdots, \ 1 \le s_0 < s_1 < s_2 < \cdots,$$

$$\alpha_{(0,b)} : \ y = b_1 x^{s_1} + b_2 x^{s_2} + \cdots, \ 1 < s_1 < s_2 < \cdots.$$

We set $G(x, y) := (f_x(x, y)^2 + f_y(x, y)^2)^{3/2}$.

By an easy computation, we have

$$f_x = 3(x - y^2)^2 + 2xy^6 - y^8,$$
$$f_y = -6y(x - y^2)^2 + 6x^2y^5 - 8xy^7 + 12y^{11},$$
$$f_{xx} = 6(x - y^2) + 2y^6,$$
$$f_{xy} = -12y(x - y^2) + 12xy^5 - 8y^7,$$
$$f_{yy} = -6(x - y^2)^2 + 24y^2(x - y^2) + 30x^2y^4 - 56xy^6 + 132y^{10}.$$

Then we have

$$\Delta_f = -(-6y(x - y^2)^2 + 6x^2y^5 - 8xy^7 + 12y^{11})^2(6(x - y^2) + 2y^6)$$
$$- (3(x - y^2)^2 + 2xy^6 - y^8)^2(-6(x - y^2)^2 + 24y^2(x - y^2) + 30x^2y^4 - 56xy^6 + 132y^{10})$$
$$+ 2(3(x - y^2)^2 + 2xy^6 - y^8)$$
$$\times (-6y(x - y^2)^2 + 6x^2y^5 - 8xy^7 + 12y^{11})(-12y(x - y^2) + 12xy^5 - 8y^7),$$

$$G = \{(3(x-y^2)^2 + 2xy^6 - y^8)^2 + (-6y(x-y^2)^2 + 6x^2y^5 - 8xy^7 + 12y^{11})^2\}^{3/2}.$$

In order to see that non concentration of curvature happens, we have to know the order of $\overline{K}_f^{\mathbb{R}}$ in y and x along α_a and $\alpha_{(0,b)}$, respectively. We first compute the order of $\overline{K}_f^{\mathbb{R}}$ along α_a. If there is no cancellation of the coefficients of the lowest order terms of $-f_y^2 f_{xx}$, $-f_x^2 f_{yy}$ and $2f_x f_y f_{xy}$, it is enough to consider the orders and coefficients of the lowest order terms of f_x, f_y, f_{xx}, f_{xy} and f_{yy}. But if the cancellation happens, we have to pay attention to more terms of them. We divide the situation into three cases. In the first two cases the cancellation of the coefficients of the lowest order terms does not happen, but such a cancellation happens in the third case.

Case (I; α_a) : $1 \leq s_0 < 2$, $a_0 \neq 0$.

Along α_a we have

$$f_x \sim 3a_0^2 y^{2s_0}, \quad f_y \sim -6a_0^2 y^{2s_0+1},$$

$$f_{xx} \sim 6a_0 y^{s_0}, \quad f_{xy} \sim -12a_0 y^{s_0+1}, \quad f_{yy} \sim 6a_0 y^{2s_0}.$$

Therefore we have

$$-f_x^2 f_{yy} \sim 54a_0^6 y^{6s_0}, \quad -f_y^2 f_{xx} \sim -216a_0^5 y^{5s_0+2}, \quad 2f_x f_y f_{xy} \sim 432a_0^5 y^{5s_0+2},$$

and

$$\Delta_f \sim 54 a_0^6 y^{6s_0}, \ G \sim 27 a_0^6 y^{6s_0}$$

along α_a. It follows that $\overline{K}_f^{\mathbb{R}} = \frac{|\Delta_f|}{G}$ tends to 2 along α_a as $y \to +0$.

Case (II; α_a) : $s_0 = 2$, $a_0 \neq 1$.

Along α_a we have

$$f_x \sim 3(a_0 - 1)^2 y^4, \quad f_y \sim -6(a_0 - 1)^2 y^5, \quad f_{xx} \sim 6(a_0 - 1)y^2,$$

$$f_{xy} \sim -12(a_0 - 1)y^3, \quad f_{yy} \sim -6(a_0 - 1)(a_0 - 5)y^4.$$

Therefore we have

$$-f_x^2 f_{yy} \sim 54(a_0 - 1)^5 (a_0 - 5)y^{12},$$
$$-f_y^2 f_{xx} \sim -216(a_0 - 1)^5 y^{12},$$
$$2 f_x f_y f_{xy} \sim 432(a_0 - 1)^5 y^{12},$$

and

$$\Delta_f \sim 54(a_0 - 1)^6 y^{12}, \ G \sim 27(a_0 - 1)^6 y^{12}$$

along α_a. It follows that $\overline{K}_f^{\mathbb{R}}$ tends to 2 along α_a as $y \to +0$.

Case (III; α_a) : $s_0 = 2$, $a_0 = 1$.

In this case the cancellation of the coefficients mentioned above happens. We set $A := a_1 y^{s_1} + a_2 y^{s_2} + \cdots$. We further divide this case into three cases.

(1) Case (III-1) : $2 < s_1 < 4$, $a_1 \neq 0$.

Along α_a we have

$$
\begin{aligned}
f_x &= 3A^2 + y^8 + HOT, \\
f_y &= -6A^2 y - 2y^9 + HOT = -2y(3A^2 + y^8) + HOT, \\
f_{xx} &= 6A + 2y^6 + HOT, \\
f_{xy} &= -12Ay + 4y^7 + HOT, \\
f_{yy} &= -6A^2 + 24Ay^2 - 26y^8 + HOT.
\end{aligned}
$$

Therefore we have

$$
\begin{aligned}
-f_x^2 f_{yy} &= (6A^2 - 24Ay^2 + 26y^8)(3A^2 + y^8)^2 + HOT, \\
-f_y^2 f_{xx} &= -(24Ay^2 + 8y^8)(3A^2 + y^8)^2 + HOT, \\
2f_x f_y f_{xy} &= (48Ay^2 - 16y^8)(3A^2 + y^8)^2 + HOT,
\end{aligned}
$$

and

$$\Delta_f \sim (6A^2 + 2y^8)(3A^2 + y^8)^2 \sim 54 a_1^6 y^{6s_1}, \ G \sim 27 a_1^6 y^{6s_1}$$

along α_a. It follows that $\overline{K}_f^{\mathbb{R}}$ tends to 2 along α_a as $y \to +0$.

(2) Case (III-2) : $s_1 = 4$, $a_1 \neq 0$.

Along α_a we have

$$f_x \sim (3a_1^2 + 1)y^8, \quad f_y \sim -2(3a_1^2 + 1)y^9, \quad f_{xx} = 6A + 2y^6 + HOT,$$

$$f_{xy} = -12Ay + 4y^7 + HOT, \quad f_{yy} = -6a_1^2y^8 + 24Ay^2 - 26y^8 + HOT.$$

Therefore we have

$$-f_x^2 f_{yy} = (6a_1^2 y^6 - 24A + 26y^6)(3a_1^2 + 1)^2 y^{18} + HOT,$$
$$-f_y^2 f_{xx} = -(24A + 8y^6)(3a_1^2 + 1)^2 y^{18} + HOT,$$
$$2f_x f_y f_{xy} = (48A - 16y^6)(3a_1^2 + 1)^2 y^{18} + HOT,$$

and

$$\Delta_f \sim 2(3a_1^2 + 1)^3 y^{24}, \quad G \sim (3a_1^2 + 1)^3 y^{24}$$

along α_a. It follows that $\overline{K}_f^{\mathbb{R}}$ tends to 2 along α_a as $y \to +0$.

(3) Case (III-3) : $s_1 > 4$.

Along α_a we have

$$f_x \sim y^8, \quad f_y \sim -2y^9, \quad f_{xx} = 6A + 2y^6 + HOT,$$

$$f_{xy} = -12Ay + 4y^7 + HOT, \quad f_{yy} = 24Ay^2 - 26y^8 + HOT.$$

Therefore we have

$$-f_x^2 f_{yy} = (-24A + 26y^6)y^{18} + HOT$$
$$-f_y^2 f_{xx} = -(24A + 8y^6)y^{18} + HOT,$$
$$2f_x f_y f_{xy} = (48A - 16y^6)y^{18} + HOT,$$

and

$$\Delta_f \sim 2y^{24}, \quad G \sim y^{24}$$

along α_a. It follows that $\overline{K}_f^{\mathbb{R}}$ tends to 2 along α_a as $y \to +0$.

We next compute the order of $\overline{K}_f^{\mathbb{R}}$ along $\alpha_{(0,b)}$. In the case where $b_1 = 0$, namely y is identically zero, we have

$$f_x = 3x^2, \quad f_y = 0, \quad f_{xx} = 6x, \quad f_{xy} = 0, \quad f_{yy} = -6x^2$$

along $\alpha_{(0,b)}$. Therefore we have

$$-f_x^2 f_{yy} = 54x^6, \quad -f_y^2 f_{xx} = 0, \quad 2f_x f_y f_{xy} = 0,$$

and

$$\Delta_f = 54x^6, \quad G = 27x^6$$

along $\alpha_{(0,b)}$. It follows that $\overline{K}_f^{\mathbb{R}}$ tends to 2 along $\alpha_{(0,b)}$ as $x \to +0$.

In the case where $b_1 \neq 0$, we have

$$f_x \sim 3x^2, \quad f_y \sim -6b_1 x^{s_1+2}, \quad f_{xx} \sim 6x, \quad f_{xy} \sim -12b_1 x^{s_1+1}, \quad f_{yy} \sim -6x^2$$

along $\alpha_{(0,b)}$. Therefore we have

$$-f_x^2 f_{yy} = 54x^6, \quad -f_y^2 f_{xx} = -216b_1^2 x^{2s_1+5}, \quad 2f_x f_y f_{xy} = 432b_1^2 x^{2s_1+5},$$

and

$$\Delta_f \sim 54x^6, \quad G \sim 27x^6$$

along $\alpha_{(0,b)}$. It follows that $\overline{K}_f^{\mathbb{R}}$ tends to 2 along $\alpha_{(0,b)}$ as $x \to +0$.

Lastly we show that f is not homogeneous-like. After taking an analytic transformation of $\mathbb{R}^2$ at $0 \in \mathbb{R}^2$, $x = X^2 + Y$, $y = Y$, f has the following form:

$$f(X,Y) = X^3 + X^2 Y^6 + X Y^8 + Y^{12}.$$

Then we consider the family of polynomials $f_t : (\mathbb{R}^2, 0) \to (\mathbb{R}, 0)$, $t \in I = [0,1]$, defined by

$$f_t(x,y) = x^3 + x^2 y^6 + t x y^8 + y^{12}.$$

For any $t \in I$, the weighted initial form of f_t with respect to the system of weights $(\frac{1}{3}, \frac{1}{12})$ is

$$g_t(x,y) = x^3 + t x y^8 + y^{12},$$

and g_t has an isolated singularity at $0 \in \mathbb{R}^2$. Therefore, by the blow-analytic triviality theorem in [4] (or [2]), we can see that f is blow-analytically equivalent to $h(x,y) = x^3 + y^{12}$. Since h clearly has a bar of height 4, it follows from Theorem 3.3 that f is not homogeneous-like. $\qquad\square$

References

1. E. García Barroso and B. Teissier, *Concentration multi-échelles de courbure dans des fibres de Milnor*, Comment. Math. Helv. **74** (1999), 398–418.
2. T. Fukui and E. Yoshinaga, *The modified analytic trivialization of family of real analytic functions*, Invent. math. **82** (1985), 467–477.
3. T. Fukui, S. Koike and T.-C. Kuo, *Blow-analytic equisingularities, properties, problems and progress*, Real Analytic and Algebraic Singularities (T. Fukuda, T. Fukui, S. Izumiya and S. Koike, ed), Pitman Research Notes in Mathematics Series, **381** (1998), pp. 8–29.

4. T. Fukui and L. Paunescu, *Modified analytic trivialization for weighted homogeneous function-germs*, J. Math. Soc. Japan **52** (2000), 433–446.
5. T. Fukui and L. Paunescu, *On blow-analytic equivalence*, Arc-Spaces and Additive Invariants in Real Algebraic Geometry, Panoramas et Syntheses **24** (2008), SMF, pp. 87–125.
6. S. Koike and A. Parusiński, *Blow-analytic equivalence of two variable real analytic function germs*, Journal of Algebraic Geometry **19** (2010), 439–472.
7. S. Koike and A. Parusiński, *Equivalence relations of two variable real analytic function germs*, Jour. Math. Soc. Japan **65** (2013), 237–276.
8. S. Koike, T.-C. Kuo and L.Paunescu, *A study of curvature using infinitesimals*, Proc. Japan Acad. Ser. A, Math. Sci. **88**, No. 5, 70–74 (2012).
9. S. Koike, T.-C. Kuo and L.Paunescu, *A'Campo curvature bumps and the Dirac phenomenon near a singular point*, arXiv:1206.0525
10. S. Koike, T.-C. Kuo and L.Paunescu, *A'Campo bumps near singular points of real analytic two variable function germs*, in preparation.
11. T.-C. Kuo, *The jet space $J^r(n,p)$*, Proceedings of the Liverpool Singularities Symposium (C. T. C. Wall, ed.), Springer Lect. Notes in Math. **192** (1971), 169–177.
12. T.-C. Kuo and Y.C. Lu, *On analytic function germs of complex variables*, Topology **16** (1977), 299–310.
13. T.-C. Kuo, *On classification of real singularities*, Invent. math. **82** (1985), 257–262.
14. T.-C. Kuo and L. Paunescu, *Equisingularity in $\mathbf{R}^2$ as Morse stability in infinitesimal calculus*, (Communicated by Heisuke Hironaka) Proc. Japan Acad. Ser. A, Math. Sci. **81**, No. 6, 115–120 (2005).
15. T.-C. Kuo and L. Paunescu, *Enriched Riemann sphere, Morse stability and equisingularity in $\mathcal{O}_2$*, Jour. London Math. Soc. **85** (2012), 382–408.
16. R. Langevin, *Courbure et singularités complexes*, Comment. Math. Helv. **54** (1979), 6–16.
17. A. Parusiński, *A criterion for the topological equivalence of two variable complex analytic function germs*, Proc. Japan Acad. Ser. A. Math. Sci. **84** No. 8 (2008), 147–150.
18. J.-J. Risler, *On the curvature of the real Milnor fiber*, Bull. London Math. Soc. **35** (2003), 445–454.
19. John A. Thorpe, *Elementary Topics in Differential Geometry*, Undergraduate Texts in Mathematics, 1979 Springer-Verlag.
20. O. Zariski, *On the topology of algebroid singularities*, Amer. Jour. Math. **54** (1932), 453–465.

Saito free divisors in four dimensional affine space and reflection groups of rank four

Jiro Sekiguchi[*]

Department of Mathematics, Tokyo University of Agriculture and Technology, Koganei, Tokyo 184-8588, Japan
sekiguti@cc.tuat.ac.jp

In this paper we collect examples of Saito free divisors in a four dimensional affine space. Some of them are already given in literatures but some are new. Typical examples are those defined as discriminant sets of real and complex reflection groups of rank four. Our interest is to study a relationship between such divisors and Saito free divisors obtained by restrictions of discriminants of polynomials to subspaces of the spaces of coefficients. In the last section, we show systems of uniformization equations with singularities along the discriminant sets of the reflection groups No. 28 and No. 31.

Keywords: free divisor, complex reflection group
AMS classification numbers: 20F65

1. Introduction

In this paper, we report the progress on the construction of examples of Saito free divisors defined as zero sets of polynomials in four variables. The author studied in detail Saito free divisors in $\mathbf{C}^3$ and related topics in a series of papers [8], [9], [10], [11], [12], [13], [3]. Compared with three dimensional case, it is difficult to construct and classify Saito free divisors in $\mathbf{C}^4$ in a systematic way. (A systematic method of constructing Saito free divisors in the three dimensional case is explained in Remark 3.1 of the main context.)

We explain an idea of finding polynomials which define Saito free divisors. Let

$$P_n(t) = t^n + x_1 t^{n-1} + x_2 t^{n-2} + \cdots + x_{n-1}t + x_n$$

[*]Partially supported by Grand-in-Aid for Scientific Research (No. 20540066, No. 2354077), Japan Society of the Promotion of Science.

be a polynomial of t. The discriminant $\Delta_n(x_1, x_2, \ldots, x_n)$ of $P_n(t)$ is regarded as a polynomial of coefficients of $P_n(t)$. Let Y be an affine subspace of $\mathbf{C}^n$ and consider the restriction $F(y_1, y_2, \ldots, y_m)$ of $\Delta_n(x_1, x_2, \ldots, x_n)$ to Y, where $y = (y_1, y_2, \ldots, y_m)$ is a coordinate on Y. Then it sometimes happens that the hypersurface $F(y_1, y_2, \ldots, y_m) = 0$ turns out to be a Saito free divisor in Y. This is a basic idea of finding Saito free divisors employed in this paper. We need some modification of this idea when we apply this idea to individual cases. Some of the discriminants of real and complex reflection groups are obtained in this manner (cf. [3]). The main result of this paper is to collect Saito free divisors in $\mathbf{C}^4$ obtained by this idea and to compare such divisors with discriminants of real and complex reflection groups of rank four. The construction of systems of uniformization equations with singularities along Saito free divisors collected in this paper is an interesting subject to attack. We only report two examples of such systems in this paper and further studies on this subject will be postponed in further consideration.

We now briefly explain the contents of this paper. In §2, we review Saito free divisors and complex reflection groups. In §3, we collect polynomials of four variables which define Saito free divisors. Some of them are discriminant sets of complex reflection groups and some are obtained as restrictions of discriminants of polynomials. In §4, we show systems of uniformization equations with singularities along the discriminant sets of the reflection groups No.28 and No.31 in the sense of [14].

In closing this introduction, we note that our computations in this paper are performed using the computer algebrac software Mathematica.

Acknowledgement: The author thanks the referee for reading the contents carefully. Following his advice, the author improved the first draft.

2. Preliminaries

2.1. *Saito free divisors*

The notion of Saito free divisors is introduced by K. Saito [6].

Let $F(x) = F(x_1, x_2, \ldots, x_n)$ be a reduced polynomial with the following conditions;

(A1) There is a vector field $E = \sum_{i=1}^n m_i x_i \partial_{x_i}$ such that $EF = dF$, where $m_1, m_2, \ldots, m_n, d$ are positive integers with $0 < m_1 \le m_2 \le \cdots \le m_n$.

(A2) There are vector fields $V^i = \sum_{j=1}^n a_{ij}(x) \partial_{x_j}$ $(i = 1, 2, \ldots, n)$ such that each $a_{ij}(x)$ is a polynomial of $x_1, x_2, \ldots, x_n$, that the determinant of the $n \times n$ matrix $(a_{ij}(x))$ coincides with $F(x)$, that $V^1 = E$, $V^i F(x) = c_i(x) F(x)$ for polynomials $c_i(x)$, that $[E, V^i] = k_i V^i$ for constants k_i and that $[V^i, V^j] \in \sum_{k=1}^n R V^k$ for any i, j, where $R = \mathbf{C}[x_1, x_2, \ldots, x_n]$.

Definition 2.1. Let $F(x)$ be a reduced polynomial. Then $D = \{x \in \mathbf{C}^n; F(x) = 0\}$ is a (weighted homogeneous) Saito free divisor if $F(x)$ satisfies conditions (A1) and (A2).

Remark 2.1. If a polynomial $F(x)$ satisfies Condition (A1), we say that $F(x)$ is weighted homogeneous with weight system $(m_1, \ldots, m_n; d)$.

Remark 2.2. A Saito free divisor is also called a logarithmic free divisor (cf. [6]).

For simplicity, we put $M = (a_{ij}(x))$ and $Der_{\mathbf{C}^n}(\log D) = \sum_{k=1}^n R V^k$. In this paper $V^1, V^2, \ldots, V^n$ are called basic vector fields for the polynomial $F(x)$ and M is called the generating matrix defining basic vector fields. It follows from the definition that

$$^t(V^1, V^2, \cdots, V^n) = M^t(\partial_{x_1}, \partial_{x_2}, \cdots, \partial_{x_n}).$$

We may assume that each $a_{ij}(x)$ is weighted homogeneous.

2.2. Irreducible real and complex reflection groups of rank four

In this subsection, we collect elementary results on irreducible complex reflection groups of rank four. A basic reference is Shephard-Todd [14].

Reflection groups treated in this subsection are real reflection groups of types A_4, B_4, D_4, F_4, H_4 and complex reflection groups of No.29, No.31, No.32 in the sense of [14].

Let G be one of the reflection groups mentioned above. Let V be a standard representation space of G over $\mathbf{C}$. (In the case of real reflection groups, we consider their complexifications.) Let P_1, P_2, P_3, P_4 be algebraically independent basic G-invariant polynomials and put $k_j = \deg_\xi(P_j)$, where $\xi = (\xi_1, \xi_2, \xi_3, \xi_4)$ are coordinates of V. We may assume that $k_1 \le k_2 \le k_3 \le k_4$. Let r be the greatest common divisor of k_1, k_2, k_3, k_4 and put $k'_j = k_j/r$ $(j = 1, 2, 3, 4)$. Since the discriminant of the group G is expressed as a polynomial of P_1, P_2, P_3, P_4, we write it by $\delta_G(P_1, P_2, P_3, P_4)$. Putting $E = \sum_{i=1}^4 k'_i P_i \partial_{P_i}$, we find that $E \delta_G = (d/r) \delta_G$, where d is the degree of δ_G.

TABLE I

	group	order	k_1, k_2, k_3, k_4	degree	(k_1', k_2', k_3', k_4')
A_4	$W(A_4)$	120	$2, 3, 4, 5$	20	$(2, 3, 4, 5)$
B_4	$W(B_4)$	384	$2, 4, 6, 8$	32	$(1, 2, 3, 4)$
D_4	$W(D_4)$	192	$2, 4, 4, 6$	24	$(1, 2, 2, 3)$
No.28	$W(F_4)$	1152	$2, 6, 8, 12$	48	$(1, 3, 4, 6)$
No.29	G_{7680}	7680	$4, 8, 12, 20$	80	$(1, 2, 3, 5)$
No.30	$W(H_4)$	14400	$2, 12, 20, 30$	120	$(1, 6, 10, 15)$
No.31	G_{46080}	46080	$8, 12, 20, 24$	120	$(2, 3, 5, 6)$
No.32	G_{155520}	155520	$12, 18, 24, 30$	120	$(2, 3, 4, 5)$

2.3. *Discriminants of reflection groups*

We retain the notation of the previous subsection. It is underlined (cf. [6], [4]) that the hypersurface $D : \delta_G(P_1, P_2, P_3, P_4) = 0$ in $\mathbf{C}^4$ is a Saito free divisor. As a consequence, if M is the generating matrix for $\delta_G(P_1, P_2, P_3, P_4)$, it follows that $\det M$ coincides with δ_G up to a non-zero constant factor. For this reason, the determination of M is important in the study of discriminants. A method how to construct M is shown in [6] for the case where G is a real reflection group and done in [4] for the case where G is a complex reflection group.

Since the construction of M is one of central subjects in our consideration, we explain briefly the method of constructing M for the case of an arbitrary reflection group. In the real case, there exists a basic G-invariant P_1 of degree two. The polynomial P_1 plays a basic role in the construction of M. On the contrary to the real case, there is no basic G-invariant of degree two in the complex case, which is one of the reasons why the argument of showing the existence of M in the real case does not go well for the complex case. Orlik and Terao overcame this difficulty and established a method of the construction of M. For the details, see Chapter 6 and Appendix B of [4].

3. Saito free divisors in $\mathbf{C}^4$

In this section, we introduce Saito free divisors in $\mathbf{C}^4$ some of which are already known and some of which are new. In Yano and Sekiguchi [15], we computed basic vector fields for discriminants of irreducible real reflection groups except those of types E_7, E_8. We collect the generating matrices which define basic vector fields for discriminants for those of types

F_4, H_4, A_4 in §3.1, §3.3, §3.5, respectively. We do not treat the cases B_4, D_4. For the readers who are interested in these cases, refer to [15]. We also collect the generating matrices in the cases of the complex reflection groups No.29, No.31 which are determined by the data in Appendix B of [4] and their explicit forms are shown in [1].

In §3.1, we treat polynomials with weight system (1,3,4,6;24). One is the discriminant of type F_4 and the other is constructed by the method explained in Introduction. Both define Saito free divisors. In §3.2, we treat polynomial of weight system (1,2,3,5;20). One is the discriminant of the complex reflection group No.29. There is another polynomial which defines Saito free divisor. In §3.3, we show the generating matrix for the discriminant of the real reflection group of type H_4. Its generating matrix is the one given in [15]. The weight system of the discriminant is (1,6,10,15;60). In §3.4, we show the generating matrix in [1] for the discriminant of the complex reflection group No.31. In this case, the discriminant is regarded as the restriction of the discriminant of the polynomial of 6th degree to a one codimensional subspace. In §3.5, we show two matrices whose determinants define Saito free divisors. The first one corresponds to the discriminant of type A_4. Weight systems of both are (2,3,4,5;20).

In the sequel, the variables will not called x_1, x_2, x_3, x_4 but rather subscripts will indicate the degree of variable or weight degree of vector fields.

Remark 3.1. In the subsequent subsections, we construct two examples of Saito free divisors in $\mathbf{C}^4$ by the method explained in Introduction (cf. §3.1, §3.2) and one example by direct computation (cf. §3.5) which are not defined by the zero sets of discriminants of reflection groups. In the three variables case, there is a systematic method of constructing Saito free divisors from isolated curve singularities (cf. [8]). We now explain this method briefly by taking a curve $-27x_3^4 + 256x_4^3 = 0$ which has E_6-singularity at the origin. Weights of x_3, x_4 are given as 3, 4, respectively. Then we consider a polynomial

$$F = 256x_4^3 - 27x_3^4 - 128x_2^2x_4^2 + (c_1x_2x_3^2 + c_2x_1^4)x_4 + c_3x_2^3x_3^2 + c_4x_2^6,$$

where c_1, c_2, c_3, c_4 are constants and x_2 is a parameter of weight 2. Then F is a weighted homogeneous polynomial of (x_2, x_3x_4) and $F = 0$ is regarded as a 1-parameter family of curves in the (x_3, x_4)-space with the parameter x_2. Computing the condition for the constants c_1, c_2, c_3, c_4 so that $F = 0$ is a Saito free divisor in $\mathbf{C}^3$, we obtain two kinds of Saito free divisors. One is defined as the zero set of the discriminant of type A_3. The other is the one obtained by M. Sato in 1970's. The author testify the method explained here to not only the cases of simple curve singularities (cf. [9]) but also some of fourteen families of exceptional singularities in the sense of Arnol'd and obtained a great number of examples of Saito free divisors including the discriminants of real and complex reflection groups of rank three (unpublished). Unfortunately the method of constructing Saito free divisors explained above is not generalized to the four variables case for the present.

3.1. The group No. 28 case (weight system $(1, 3, 4, 6; 24)$)

The group No.28 is identified with the Weyl group $W(F_4)$ of type F_4. The discriminant of $W(F_4)$ is computed in [15] and its generating matrix $M(F_4)$ is given as follows:

$$
M(F_4) = \begin{pmatrix}
y_2 & 3y_6 & 4y_8 & 6y_{12} \\
3y_6 & -15y_2y_8 + \frac{3}{2}y_2^5 & -18y_{12} - 6y_2^3y_6 & 4y_8^2 - 4y_2^4y_8 + 4y_2^2y_6^2 \\
4y_8 & -18y_{12} - 6y_2^3y_6 & -7v_2^3v_8 + 7v_2v_6^2 + \frac{1}{2}v_2^7 & 9v_2^2v_6v_8 - v_6^3 - \frac{3}{2}v_2^6v_6 \\
6y_{12} & 4v_8^2 - 4v_2^4v_8 + 4v_2^2v_6^2 & 9v_2^2v_6v_8 - v_6^3 - \frac{3}{2}v_2^6v_6 & \frac{11}{3}v_2^3v_8^2 - \frac{11}{3}v_2v_6^2v_8 \\
& & & +\frac{11}{6}v_2^5v_6^2 + \frac{1}{12}v_2^{11}
\end{pmatrix}
$$

Here y_2, y_6, y_8, y_{12} are basic $W(F_4)$-invariants.

In this case, we consider the polynomial of t defined by

$$
27t^8 + (4x_1^3 + 54x_3)t^6 + (54x_6 + 27x_3^2 + 12x_1^2x_4)t^4
$$
$$
+ (54x_6x_3 + 12x_1x_4^2)t^2 + 4x_4^3 + 27x_6^2.
$$

By direct computation we find that its discriminant coincides with

$$
(x_6x_1^2 - x_1x_3x_4 + x_4^2)^6(27x_6^2 + 4x_4^3)
$$
$$
\times(432x_6^2 - x_6x_1^6 - 36x_6x_1^3x_3 - 216x_6x_3^3 + x_1^3x_3^3 + 27x_3^4
$$
$$
+72x_6x_1^2x_4 + x_1^5x_3x_4 + 30x_1^2x_3^2x_4 - x_1^4x_4^2 - 96x_1x_3x_4^2 + 64x_4^3)^2
$$

up to a constant factor. Then

$$
(27x_6^2 + 4x_4^3)
$$
$$
\times(432x_6^2 - x_6x_1^6 - 36x_6x_1^3x_3 - 216x_6x_3^3 + x_1^3x_3^3 + 27x_3^4
$$
$$
+72x_6x_1^2x_4 + x_1^5x_3x_4 + 30x_1^2x_3^2x_4 - x_1^4x_4^2 - 96x_1x_3x_4^2 + 64x_4^3)
$$

coincides with the discriminant $\det M(F_4)$ by the correspondence

$$
\begin{aligned}
x_1 &= y_2, \\
x_3 &= \tfrac{1}{72}(-3y_2^3 + 2y_6), \\
x_4 &= \tfrac{1}{576}(-y_2^4 + 4y_2y_6 + 4y_8), \\
x_6 &= \tfrac{1}{10368}(-6y_{12} - y_2^3y_6 + y_6^2 - 3y_2^2y_8).
\end{aligned}
$$

Remark 3.2. The differences of the indices of x_1, x_3, x_4, x_6 and y_2, y_6, y_8, y_{12} are confusing. This is caused by the notation of basic $W(F_4)$-invariants employed in [15].

We are now going to find a polynomial of four variables with weight system $(1,3,4,6;24)$ which defines a Saito free divisor not equal to that defined by the discriminant of $W(F_4)$. Let

$$P_6(t) = t^6 + x_2 t^4 + x_3 t^3 + x_4 t^2 + x_5 t + x_6$$

be a polynomial of t and let $\Delta_6(x_2, x_3, x_4, x_5, x_6)$ be the discriminant of $P_6(t)$. We treat a polynomial $f(x_1, x_3, x_4, x_6)$ obtained by restricting $\Delta_6(x_2, x_3, x_4, x_5, x_6)$ to the subspace defined by

$$x_5 = a_1 x_4 x_1 + a_2 x_3 x_1^2 + a_3 x_1^4,$$
$$x_2 = b_1 x_1^2,$$

where a_1, a_2, a_3, b_1 are constants. We first consider the condition that $f(x_1, x_3, x_4, x_6)$ is divided by

$$x_6 + c_1 x_4 x_1^2 + c_2 x_3^2 + c_3 x_3 x_1^3 + c_4 x_1^6.$$

By direct computation, we find that if

$$a_2 = -\tfrac{3}{4} a_1^2,$$
$$a_3 = \tfrac{1}{16} a_1^3(-5a_1 + 3a_1^2 + 8b_1)$$

and

$$c_1 = -\tfrac{1}{4} a_1^2,$$
$$c_2 = 0,$$
$$c_3 = \tfrac{1}{4} a_1^3,$$
$$c_4 = -\tfrac{1}{64} a_1^4(-8a_1 + 5a_1^2 + 12b_1,$$

then $f(x_1, x_3, x_4, x_6)$ is reducible. As a consequence, if $a_2, a_3, c_1, c_2, c_3, c_4$ satisfy the relations above, there is a polynomial $f_0(x_1, x_3, x_4, x_6)$ such that

$$f(x_1, x_3, x_4, x_6) = \{64 x_6 - a_1^4(-8a_1 + 5a_1^2 + 12b_1)x_1^6 + 16a_1^3 x_1^3 x_3 - 16a_1^2 x_1^2 x_4\}$$
$$\times f_0(x_1, x_3, x_4, x_6).$$

Next we consider the condition that

$$f_0(x_1, x_3, x_4, x_6) = 0$$

is a Saito free divisor. By direct computation, we conclude that if

$$a_1 = \frac{1}{6}(2 - p_0), \quad b_1 = \frac{1}{24}(10 - p_0^2)$$

for a non-zero constant p_0, then

$$f_0(x_1, x_3, x_4, x_6) = 0$$

is a Saito free divisor. It can be shown that the hypersurface

$$f_0(x_1, x_3, x_4, x_6) = 0$$

is isomorphic to each other for arbitrary $p_0 (\neq 0)$. In particular, in the case $p_0 = -2$, $f_0(x_1, x_3, x_4, x_6)$ is equal to the polynomial $f_{No.28}(x_1, x_3, x_4, x_6)$ defined as the determinant of the following matrix up to a non-zero constant factor:

$$
\begin{pmatrix}
x_1 & 3x_3 & 4x_4 & 6x_6 \\
54x_3 & x_1(5x_1x_3 - 22x_4) & \begin{matrix} 7x_1^3x_3 + 36x_3^2 \\ -20x_1^2x_4 - 36x_6 \end{matrix} & \begin{matrix} -4x_1x_3x_4 + 8x_4^2 \\ -12x_1^2x_6 \end{matrix} \\
16x_4 & 30x_3^2 - 24x_6 & \begin{matrix} -7x_1x_3^2 + 52x_3x_4 \\ -12x_1x_6 \end{matrix} & \begin{matrix} x_1^3x_3^2 - 4x_1^2x_3x_4 \\ +4x_1x_4^2 + 48x_3x_6 \end{matrix} \\
216x_6 & \begin{matrix} -19x_1^2x_3^2 + 2x_1x_3x_4 \\ +48x_4^2 \end{matrix} & \begin{matrix} 22x_1^3x_3^2 + 9x_3^3 \\ -116x_1^2x_3x_4 + 132x_1x_4^2 \\ +72x_3x_6 \end{matrix} & \begin{matrix} -3x_1^2x_3^3 + 8x_1x_3^2x_4 \\ -4x_3x_4^2 - 84x_1^2x_3x_6 \\ +144x_1x_4x_6 \end{matrix}
\end{pmatrix}.
$$

As a consequence, we have the following proposition:

Proposition 3.1. *There are at least two polynomials with weight system (1,3,4,6;24) of four variables which define Saito free divisors. One is the discriminant of $W(F_4)$ and the other is $f_{No.28}(x_1, x_3, x_4, x_6)$.*

Remark 3.3. In the occasion of the workshop of Japan-Russia bilateral program held at Krasnoyarsk University in August, 2006, the author gave a talk on Saito free divisors and showed that $f_{No.28}(x_1, x_3, x_4, x_6) = 0$ is a Saito free divisor.

3.2. The group No. 29 case (weight system $(1, 2, 3, 5; 20)$)

The generating matrix for the discriminant of the group No.29 is obtained in [1]. Its concrete form is given by

$$
M_{29} =
\begin{pmatrix}
320x & 640y & 960z & 1600t \\
640y^2 & 1280(3200t + xy^2 + 176yz) & -640x(20t + yz) & -640(txy + 80tz + 2yz^2) \\
2xy + 960z & 4x^2y + 64xz & 200t + 3x^2z - 5yz & 8tx^2 - 10ty - 4xz^2 \\
1600t + 8yz & -640tx + 16xyz + 1536z^2 & 10tx^2 - 10ty - 8xz^2 & 72txz - 96z^3
\end{pmatrix}.
$$

There is a relationship between the determinant of M_{29} which is equal to the discriminant of the group No.29 and the discriminant of a polynomial of fifth degree which we are going to explain. The discriminant of

$$T^5 + x_1 T^4 + x_2 T^3 + x_3 T^2 + \frac{x_2^2}{4} T + x_5$$

coincides with the determinant of the above matrix by the coordinate transformation

$$
\begin{aligned}
x &= 4x_1, \\
y &= -8x_2, \\
z &= \tfrac{1}{2}x_1 x_2 - \tfrac{1}{2}x_3, \\
t &= \tfrac{1}{8}x_1 x_2^2 - \tfrac{1}{4}x_2 x_3 + \tfrac{1}{2}x_5.
\end{aligned}
$$

There is another Saito free divisor which is the zero locus of the polynomial of x_1, x_2, x_3, x_5 obtained by the restriction of the discriminant $\Delta_5(x_1, x_2, x_3, x_4, x_5)$ of the polynomial

$$T^5 + x_1 T^4 + x_2 T^3 + x_3 T^2 + x_4 T + x_5$$

to a hypersurface of the $(x_1, x_2, x_3, x_4, x_5)$-space defined by

$$x_4 = a_1 x_3 x_1 + a_2 x_2^2 + a_3 x_2 x_1^2 + a_4 x_1^4$$

for some constants a_1, a_2, a_3, a_4.

By direct computation, we find that if $f(x_1, x_2, x_3, x_5)$ is the polynomial of (x_1, x_2, x_3, x_5) obtained by the substitution

$$x_4 = -\frac{(2+c)}{2000} x_1 \{(-6+c)(2+c)^2 x_1^3 + 60(2+c)x_1 x_2 - 400x_3\}$$

to $\Delta_5(x_1, x_2, x_3, x_4, x_5)$, then $f(x_1, x_2, x_3, x_5) = 0$ is a Saito free divisor, where c is a non-zero constant. To show its concrete form, it is better to change the coordinate. For this purpose, we introduce a coordinate transformation of the (x_1, x_2, x_3, x_5)-space to the (y_1, y_2, y_3, y_5)-space defined by

$$
\begin{aligned}
x_1 &= -\tfrac{6}{5c} y_1, \\
x_2 &= -\tfrac{1}{500c^2}(-288y_1^2 + 27c^2 y_1^2 - 500c^2 y_2), \\
x_3 &= \tfrac{1}{12500c^3}(-1728y_1^3 + 486c^2 y_1^3 + 27c^3 y_1^3 - 9000c^2 y_1 y_2 \\
&\qquad -4500c^3 y_1 y_2 + 12500c^3 y_3),
\end{aligned}
$$

$$x_5 = -\frac{1}{39062500c^5}(31104y_1^5 - 29160c^2y_1^5 - 4860c^3y_1^5 + 7290c^4y_1^5 + 2187c^5y_1^5$$
$$+540000c^2y_1^3y_2 + 810000c^3y_1^3y_2 + 405000c^4y_1^3y_2 + 67500c^5y_1^3y_2$$
$$-2250000c^3y_1^2y_3 - 2250000c^4y_1^2y_3 - 562500c^5y_1^2y_3 - 39062500c^5y_5).$$

Regarding $f(x_1, x_2, x_3, x_5)$ as a polynomial of (y_1, y_2, y_3, y_5), its generating matrix is given by

$$
\begin{pmatrix}
y_1 & 2y_2 & 3y_3 & 5y_5 \\
& & & \\
550000y_2^2 \ \ 2\begin{pmatrix} -3159y_1^3y_2 \\ -84600y_1y_2^2 \\ +17145y_1^2y_3 \\ +202500y_2y_3 \\ -687500y_5 \end{pmatrix} & 3\begin{pmatrix} 1323y_1^3y_3 \\ -15600y_1y_2y_3 \\ -45000y_3^2 \\ -205000y_1y_5 \end{pmatrix} & y_5\begin{pmatrix} 11907y_1^3 \\ -775000y_3 \end{pmatrix} \\
& & & \\
2500y_3 \ \ 2\begin{pmatrix} 72y_1^2y_2 \\ +800y_2^2 \\ -285y_1y_3 \end{pmatrix} & \begin{pmatrix} -27y_1^2y_3 \\ +2900y_2y_3 \\ +2500y_5 \end{pmatrix} & y_5\begin{pmatrix} -81y_1^2 \\ +5500y_2 \end{pmatrix} \\
& & & \\
2812500y_5 \ \ 10\begin{pmatrix} 1440y_1^2y_2^2 \\ +16000y_2^3 \\ +108y_1^3y_3 \\ -40500y_3^2 \\ +14375y_1y_5 \end{pmatrix} & 3\begin{pmatrix} 108y_1^4y_3 \\ +1500y_1^2y_2y_3 \\ +110000y_2^2y_3 \\ -31500y_1y_3^2 \\ +10775y_1^2y_5 \\ +487500y_2y_5 \end{pmatrix} & 4y_5\begin{pmatrix} 243y_1^4 \\ +3375y_1^2y_2 \\ +167500y_2^2 \\ -68875y_1y_3 \end{pmatrix}
\end{pmatrix}.
$$

In fact it is provable that the determinant $f_{No.29}(y_1, y_2, y_3, y_5)$ of this matrix defines a Saito free divisor in the (y_1, y_2, y_3, y_5)-space. Note that $f_{No.29}(y_1, y_2, y_3, y_5)$ has a factor y_5.

As a consequence we obtain the following proposition.

Proposition 3.2. *There are at least two polynomials with weight system $(1,2,3,5;20)$ of four variables which define Saito free divisors. One is the discriminant of the group No.29 and the other is $f_{No.29}(y_1, y_2, y_3, y_5)$.*

3.3. The group No. 30 case (weight system $(1, 6, 10, 15; 60)$)

The group No.30 is identified with the real reflection group $W(H_4)$ of type H_4. The discriminant of $W(H_4)$ and its generating matrix is obtained in [15]. In particular the generating matrix is given as follows:

$$M(H_4) = \begin{pmatrix} 2x_2 & 12x_{12} & 20x_{20} & 30x_{30} \\[6pt] 12x_{12} & \begin{matrix} 22x_2x_{20} \\ -4x_2^5x_{12} \end{matrix} & \begin{matrix} -15x_{30} \\ -12x_2^5x_{20} \\ +4x_2^3x_{12}^2 \end{matrix} & \begin{matrix} -20x_2^5x_{30} \\ -40x_{20}^2 \\ -8x_2^4x_{12}x_{20} \\ -\frac{8}{3}x_2^2x_{12}^3 \end{matrix} \\[6pt] 20x_{20} & \begin{matrix} -15x_{30} \\ -12x_2^5x_{20} \\ +4x_2^3x_{12}^2 \end{matrix} & \begin{matrix} 10x_2^4x_{30} \\ -28x_2^3x_{12}x_{20} \\ +\frac{76}{3}x_2x_{12}^3 \end{matrix} & \begin{matrix} -60x_2^3x_{12}x_{30} \\ +24x_2^4x_{20}^2 \\ -56x_2^2x_{12}^2x_{20} \\ -16x_{12}^4 \end{matrix} \\[6pt] 30x_{30} & \begin{matrix} -20x_2^5x_{30} \\ -40x_{20}^2 \\ -8x_2^4x_{12}x_{20} \\ -\frac{8}{3}x_2^2x_{12}^3r \end{matrix} & \begin{matrix} -60x_2^3x_{12}x_{30} \\ +24x_2^4x_{20}^2 \\ -56x_2^2x_{12}^2x_{20} \\ -16x_{12}^4 \end{matrix} & \begin{matrix} 80x_2^4x_{20}x_{30} \\ +120x_2^2x_{12}^2x_{30} \\ +48x_2^3x_{12}x_{20}^2 \\ +\frac{464}{3}x_2x_{12}^3x_{20} \end{matrix} \end{pmatrix}.$$

We do not check whether the discriminant of $W(H_4)$ is identified with a polynomial which is obtained as a restriction of the discriminant of a polynomial of degree 15 or not because of the difficulty of the computation.

3.4. The group No. 31 case (weight system $(2, 3, 5, 6; 24)$)

The defining matrix of the discriminant of the group No.31 is obtained in [1]. Its concrete form is shown as follows:

$$M_{31} = \begin{pmatrix} 2160x & 3240y & 5400z & 6480t \\[4pt] 3240ty & \begin{matrix} 4860tx^2 \\ -26244000z^2 \end{matrix} & -9720t^2 + 16200xz^2 & \begin{matrix} -11340txz \\ -16200yz^2 \end{matrix} \\[4pt] 2xy + 5400z & -9720t + 3x^3 & -tx - 5yz & -5ty - 2x^2z \\[4pt] 6480t - 2y^2 & \begin{matrix} -3x^2y \\ -11340xz \end{matrix} & ty + 5x^2z & \begin{matrix} 5tx^2 + 2xyz \\ +5400z^2 \end{matrix} \end{pmatrix}.$$

There is a relationship between the determinant of M_{31} and the discriminant of a polynomial of sixth degree which we are going to explain. Let

$$P_6(T) = T^6 + x_2T^4 + x_3T^3 + x_4T^2 + x_5T + x_6$$

be a polynomial of t and let $\Delta_6(x_2, x_3, x_4, x_5, x_6)$ be the discriminant of $P_6(T)$. Consider the restriction of $\Delta_6(x_2, x_3, x_4, x_5, x_6)$ to the subspace defined by $x_4 = px_2^2$ for a constant p and put

$$f(x_2, x_3, x_5, x_6; p) = \Delta_6(x_2, x_3, px_2^2, x_5, x_6).$$

Then it is possible to show that the hypersurface $f(x_2, x_3, x_5, x_6; p) = 0$ in the (x_2, x_3, x_5, x_6)-space is a Saito free divisor if and only if $p = \frac{1}{4}$. Moreover $f(x_2, x_3, x_5, x_6; \frac{1}{4})$ coincides with the determinant of M_{31} up to a constant factor by the coordinate change:

$$
\begin{aligned}
x &= -6x_2, \\
y &= -54x_3, \\
z &= -\tfrac{1}{4}x_2 x_3 + \tfrac{1}{2}x_5, \\
t &= \tfrac{3}{4}x_3^2 - 3x_6.
\end{aligned}
$$

In the next section, we will introduce systems of uniformization equations with singularities along the hypersurface $f(x_2, x_3, x_5, x_6; \frac{1}{4}) = 0$ and discuss some of its properties.

3.5. The group No. 32 case (weight system $(2, 3, 4, 5; 20)$)

The basic vector fields for the discriminant of the group No.32 are identified with those of the Weyl group $W(A_4)$ of type A_4 by choosing the basic invariants appropriately. The generating matrix for the case $W(A_4)$ is, for example, given in [15]. Its concrete form is shown as follows:

$$
\begin{pmatrix}
2x_2 & 3x_3 & 4x_4 & 5x_5 \\
3x_3 & -\frac{2}{5}(3x_2^2 - 10x_4) & \frac{1}{5}(-4x_2 x_3 + 25x_5) & -\frac{2}{5}x_2 x_4 \\
4x_4 & \frac{1}{5}(-4x_2 x_3 + 25x_5) & -\frac{2}{5}(3x_3^2 - 5x_2 x_4) & -\frac{2}{5}(x_3 x_4 - 5x_2 x_5) \\
5x_5 & -\frac{2}{5}x_2 x_4 & -\frac{2}{5}(x_3 x_4 - 5x_2 x_5) & -\frac{2}{5}(2x_4^2 - 5x_3 x_5)
\end{pmatrix}.
$$

The weight system of the discriminant in this case is $(2,3,4,5;20)$. There is another polynomial of weight system $(2,3,4,5;20)$ which defines a Saito free divisor. Its generating matrix is shown as follows (cf. [9]):

$$
M =
\begin{pmatrix}
2x_2 & 3x_3 & 4x_4 & 5x_5 \\
3x_3 & x_2^2 + 4x_4 & \frac{5}{4}x_5 & \frac{3}{10}x_2 x_4 \\
4x_4 & 5x_5 & -\frac{5}{8}x_2 x_4 & -\frac{5}{8}x_2 x_5 \\
5x_5 & 2x_2 x_4 & \frac{15}{16}x_3 x_4 & \frac{3}{5}x_4^2 + \frac{15}{16}x_3 x_5
\end{pmatrix}.
$$

4. Systems of uniformization equations

If one obtains a Saito free divisor, the next question is to ask the existence of systems of uniformization equations with singularities along it. In [13], the author treated this question for the case of the reflection group of type

D_4. In this section, we treat the cases of the groups No.28 and No.31. In particular we show examples of systems of uniformization equations with singularities along the discriminant sets for these two groups. One of our interests is to study the properties of the holonomic systems constructed in this section. We hope to discuss this subject as well as the construction of the systems of uniformization equations for the remaining cases in a future.

In the sequel, the variables will not called x_1, x_2, x_3, x_4 and the basic vector fields will not to be called V^1, V^2, V^3, V^4, but rather subscripts will indicate the degree of variable or weight degree of vector fields.

4.1. The case No. 28

We replace the variables y_2, y_6, y_8, y_{12} in §3.1 with x_1, x_3, x_4, x_6, respectively and modify the generating matrix $M(F_4)$ with $M'(F_4)$ defined by

$$M'(F_4) = \begin{pmatrix} x_1 & 3x_3 & 4x_4 & 6x_6 \\ 3x_3 & \frac{3x_1}{2}(x_1^4 - 10x_4) & -6x_1^3x_3 - 18x_6 & 4x_1^2x_3^2 - 4x_1^4x_4 + 4x_4^2 \\ \frac{1}{2}x_1^4 + 4x_4 & -\frac{9}{2}x_1^3x_3 & \frac{1}{2}x_1^7 + 7x_1x_3^2 & -\frac{3}{2}x_1^6x_3 \\ & -18x_6 & -5x_1^3x_4 & -x_3^3 + 9x_1^2x_3x_4 + 3x_1^3x_6 \\ & & & \frac{1}{12}x_1^{11} \\ -\frac{1}{2}x_1^3x_3 + 6x_6 & \frac{5}{2}x_1^2x_3^2 & -\frac{3}{2}x_1^6x_3 & +\frac{11}{6}x_1^5x_3^2 \\ & -4x_1^4x_4 & -x_3^3 + 7x_1^2x_3x_4 & -\frac{11}{3}x_1x_3^2x_4 \\ & +4x_4^2 & & +\frac{11}{3}x_1^3x_4^2 \\ & & & -3x_1^2x_3x_6 \end{pmatrix}.$$

Let $f_0 = \det(M'(F_4))$ be the polynomial which defines a Saito free divisor. Define vector fields V^0, V^2, V^3, V^5 from $M'(F_4)$. Then

$$V^0 f_0 = 24f_0, \quad V^j f_0 = 0 \ (j = 2, 3, 5),$$

$$[V^2, V^3] = -2x_1^3V^2 - 3V^5,$$
$$[V^2, V^5] = 4x_1^2x_3V^2 + (3x_4 - \tfrac{3}{2}x_1^4)V^3,$$
$$[V^3, V^5] = (2x_1^2x_4 - \tfrac{2}{3}x_3^2 - \tfrac{1}{3}x_1^6)V^2 + \tfrac{7}{2}x_1^2x_3V^3 + \tfrac{5}{2}x_1^3V^5.$$

In this case $\mathcal{M}(r_1, r_2, r_3, s)$ $(r_1, r_2, r_3, s \in \mathbf{C})$ defined below is a system of uniformization equations with singularities along the hypersurface $f_0 = 0$.

$$V^0 u = su,$$

$$V^2V^2 = \left[\begin{array}{c} \frac{1}{3}\{(6r_2 + r_2^2 - 3r_1r_3)x_1^4 + (18r_1 + 3r_1r_2 - r_2r_3)x_1x_3 \\ -r_2(6 + r_2)x_4\} + (r_1 + r_3)x_1^2V^2 + r_2x_1V^3 \end{array} \right] u,$$

$$V^3V^3u = \left[\begin{array}{l} \frac{1}{36}\left\{\begin{array}{l} 3(-36 + r_2^2 - 4r_1r_3 + 4r_3^2)x_1^6 + 24r_2r_3x_1^3x_3 \\ +4(-18 + 3r_2 + r_2^2 - 6r_1r_3 + 2r_3^2)x_3^2 \\ -12(18 + 9r_2 + r_2^2 + 2r_1r_3 - 2r_3^2)x_1^2x_4 \\ +72(6r_1 + r_1r_2 - r_2r_3)x_6 \end{array}\right\} \\ +\frac{1}{3}\{(r_1 - r_3)(x_1^4 + 4x_4) - 2(-3 + r_2)x_1x_3\}V^2 \\ +\frac{1}{6}(-9x_1^3 - 2(3r_1 + r_3)x_3)V^3 + (r_1 - r_3)x_1V^5 \end{array}\right]u,$$

$$V^5V^5u = \left[\begin{array}{l} \frac{1}{108}\left\{\begin{array}{l} (-54 - 3r_2 + r_2^2 - 6r_1r_3 + 6r_3^2)(x_1^{10} - 4x_3^2x_4) \\ +4r_2r_3(3x_1^7x_3 + 2x_1x_3^3) \\ +9(-36 + r_2^2 - 4r_1r_3 + 4r_3^2)x_1^4x_3^2 \\ -4(54 + 15r_2 + r_2^2 + 6r_1r_3 - 6r_3^2)x_1^6x_4 \\ -8(18r_1 + 3r_1r_2 - r_2r_3)x_1^3x_3x_4 \\ +12(-18 + 3r_2 + r_2^2 - 6r_1r_3 + 2r_3^2)x_1^2x_4^2 \\ +72(18 + 3r_2 + 2r_1r_3 - 2r_3^2)x_1x_3x_6 \\ -72(6r_1 + r_1r_2 - r_2r_3)x_4x_6 \end{array}\right\} \\ +\frac{1}{18}(x_1^4 - 2x_4)\left\{\begin{array}{l} (r_1 - r_3)(x_1^4 + 4x_4) \\ -2(-9 + r_2)x_1x_3 \end{array}\right\}V^2 \\ +\frac{1}{18}\left\{\begin{array}{l} 3(r_1 - r_3)(-3x_1^4x_3 + 2x_3x_4 + 12x_1x_6) \\ +2(-9 + r_2)(x_1x_3^2 - 2x_1^3x_4) \end{array}\right\}V^3 \\ -\frac{1}{6}x_1\left\{\begin{array}{l} (r_1 - r_3)x_1^4 - (-3 + 2r_2)x_1x_3 \\ +2(5r_1 - r_3)x_4 \end{array}\right\}V^5 \end{array}\right]u,$$

$$(V^2V^3 + V^3V^2)u = \left[\begin{array}{l} \frac{1}{3}x_1\left\{\begin{array}{l} (6r_1 + r_1r_2 - 2r_2r_3)x_1^4 \\ -(18r_2 + 3r_2^2 - 4r_1r_3)x_1x_3 \\ +4(12r_1 + 2r_1r_2 - r_2r_3)x_4 \end{array}\right\} \\ +\frac{1}{3}\{(-6 + r_2)x_1^3 - 4r_3x_3\}V^2 \\ +2r_1x_1^2V^3 + (-3 + 2r_2)V^5 \end{array}\right]u,$$

$$(V^2V^5 + V^5V^2)u = \left[\begin{array}{l} \frac{1}{18}\left\{\begin{array}{l} r_2(6 + r_2)(x_1^7 + 4x_1x_3^2) \\ -2(18r_1 + 3r_1r_2 - 8r_2r_3)x_1^4x_3 \\ -2(30r_2 + 5r_2^2 - 12r_1r_3)x_1^3x_4 \\ -8r_2r_3x_3x_4 + 72(6r_1 + r_1r_2 - r_2r_3)x_1x_6 \end{array}\right\} \\ -\frac{1}{3}x_1\{(-12 + r_2)x_1x_3 + 4r_3x_4\}V^2 \\ +\frac{1}{6}(-9 + 2r_2)(x_1^4 - 2x_4)V^3 + 2r_1x_1^2V^5 \end{array}\right]u,$$

$$(V^3V^5 + V^5V^3)u = \begin{bmatrix} \frac{1}{18}\left\{\begin{array}{c} (6r_1 + r_1r_2 - 3r_2r_3)x_1^8 \\ -(-144 + 6r_2 + 5r_2^2 - 16r_1r_3 + 16r_3^2)x_1^5x_3 \\ 2(18r_1 + 3r_1r_2 + r_2r_3)(x_1^4x_4 - x_1^2x_3^2) \\ +8(-18 + 3r_2 + r_2^2 - 4r_1r_3 + 2r_3^2)x_1x_3x_4 \\ -16(6r_1 + r_1r_2 - r_2r_3)x_4^2 \\ -36(18 + 3r_2 + 2r_1r_3 - 2r_3^2)x_1^2x_6 \end{array}\right\} \\ \frac{1}{9}\left\{\begin{array}{c} (-6 + r_2)(x_1^6 + 2x_3^2 - 6x_1^2x_4) \\ +6(r_1 - r_3)(6x_6 - x_1^3x_3) \end{array}\right\}V^2 \\ +\frac{1}{6}x_1\left\{\begin{array}{c} 2(r_1 - r_3)x_1^4 - (-27 + 2r_2)x_1x_3 \\ -4(r_1 + r_3)x_4 \end{array}\right\}V^3 \\ +\frac{1}{6}\{-(-3 + 2r_2)x_1^3 - 4(3r_1 - r_3)x_3\}V^5 \end{bmatrix} u.$$

The system $\mathcal{M}(r_1, r_2, r_3, s)$ sometimes has a non-trivial quotient $\mathcal{D}$-module. By direct computation we find the following:

(1) $\mathcal{M}(r_1, 0, r_3, s)$ has a quotient added by a differential equation $V^2u = r_1x_1^2u$.

(2) $\mathcal{M}(r_1, 3, r_3, s)$ has a quotient added by a differential equation

$$V^3u = \left\{\frac{1}{3}(r_1 - r_3)x_1V^2 + \frac{1}{3}(-3r_1 + r_3)x_3 + \frac{1}{6}(-9 - 2r_1^2 + 2r_1r_3)x_1^3\right\}u.$$

(3) $\mathcal{M}(r_1, 9, r_3, s)$ has a quotient added by a differential equation

$$V^5u = \left\{-\frac{1}{54}(r_1 - r_3)((r_1 - r_3)x_1^3 + 6x_3)V^2 + \frac{1}{6}(r_1 - r_3)V^3 \right.$$
$$-\frac{1}{3}(5r_1 - 3r_3)x_1x_4 + \frac{1}{18}(45 + (r_1 - r_3)(5r_1 - r_3))x_1^2x_3$$
$$\left.+ \frac{1}{108}(r_1 - r_3)(27 + 2r_1^2 - 2r_1r_3)x_1^5\right\}u.$$

4.2. The case No. 31

Let M be the matrix with polynomial entries defined by

$$\begin{pmatrix} 2x_2 & 3x_3 & 5x_5 & 6x_6 \\ 3\begin{pmatrix}-7x_2x_3\\+50x_5\end{pmatrix} & 2\begin{pmatrix}-2x_2^3\\-9x_3^2\\+90x_6\end{pmatrix} & 2\begin{pmatrix}-x_4^2\\+15x_3x_5\\+51x_2x_6\end{pmatrix} & 2\begin{pmatrix}-2x_2^2x_5\\+27x_3x_6\end{pmatrix} \\ \begin{array}{c}-8x_2^3\\+36x_3^2\\+720x_6\end{array} & 2x_2\begin{pmatrix}17x_2x_3\\-70x_5\end{pmatrix} & \begin{array}{c}3x_2^3x_3\\-50x_2^2x_5\\+432x_3x_6\end{array} & \begin{array}{c}6x_2x_3x_5\\-100x_5^2\\+16x_2^2x_6\end{array} \\ \begin{array}{c}54x_2^3x_3\\-324x_3^3\\-420x_2^2x_5\\+6480x_3x_6\end{array} & \begin{array}{c}8x_2^5 - 360x_2^2x_3^2\\+1980x_2x_3x_5\\-3000x_5^2\\-600x_2^2x_6\end{array} & \begin{array}{c}4x_2^6 - 27x_2^3x_3^2\\+90x_2^2x_3x_5\\-600x_2x_5^2\\-324x_2^3x_6\\+1512x_3^2x_6\\+4320x_6^2\end{array} & \begin{array}{c}8x_2^4x_5\\-54x_2x_3^2x_5\\-300x_3x_5^2\\+468x_2^2x_3x_6\\-1200x_2x_5x_6\end{array} \end{pmatrix}$$

and let V^j $(j = 0, 3, 4, 7)$ be vector fields on (x_2, x_3, x_5, x_6)-space defined by

$$^t(V^0, 150V^3, 720V^4, 108000V^7) = M \cdot {}^t(\partial_{x_2}, \partial_{x_3}, \partial_{x_5}, \partial_{x_6}).$$

Then $f_0 = \det M$ is regarded as the discriminant of the group No.31 by the correspondence explained in the previous section. A simple computation shows the following identities:

$$[V^0, V^j] = jV^j \quad (j = 3, 4, 7),$$
$$[V^3, V^4] = -\tfrac{1}{72}x_2^2 V^3 - \tfrac{1}{25}x_3 V^4 + V^7,$$
$$[V^3, V^7] = -\tfrac{11}{1800}x_2 x_5 V^3 + \tfrac{1}{1875}(-2x_3^3 + 9x_3^2 + 60x_6)V^4 - \tfrac{7}{50}x_3 V^7,$$
$$[V^4, V^7] = \tfrac{1}{14400}(x_2^4 + 6x_2 x_3^2 - 40x_3 x_5 - 40x_2 x_6)V^3 - \tfrac{1}{180}x_2 x_5 V^4 + \tfrac{7}{360}x_2^2 V^7,$$

$$V^0 f_0 = 30 f_0, \quad V^3 f_0 = 0, \quad V^4 f_0 = 0, \quad V^7 f_0 = 0.$$

These actually show that the hypersurface $f_0 = 0$ is a Saito free divisor in the (x_2, x_3, x_5, x_6)-space.

In this case $\mathcal{M}(r, s)$ $(r, s \in \mathbf{C})$ defined below is a system of uniformization equations with singularities along the hypersurface $f_0 = 0$.

$$V^0 u = su$$

$$V^3 V^3 u = \left[\begin{array}{l} -\frac{r-2}{3750}\{2rx_2^3 - 3(2r+5)x_3^2 - 30(r-2)x_6\} \\ +\frac{3}{50}x_3 V^3 + \frac{12(r+1)}{125}x_2 V^4 \end{array} \right] u,$$

$$V^4 V^4 u = \left[\begin{array}{l} \frac{r-2}{32400}\{rx_2^4 + 3(r+5)x_2 x_3^2 - 45(r+2)x_3 x_5 \\ -30(r-4)x_2 x_6\} \\ -\frac{r+2}{432}(x_2 x_3 + 10x_5)V^3 - \frac{1}{90}x_2^2 V^4 \end{array} \right] u,$$

$$V^7 V^7 u = \left[\begin{array}{l} -\frac{r-2}{2^8 \cdot 3^5 \cdot 5^6} \left(\begin{array}{l} 4rx_2^7 - 3(40 + 17r)x_2^4 x_3^2 \\ -324(5+r)x_2 x_3^4 + 120(-1+r)x_2^3 x_3 x_5 \\ +180(44+7r)x_3^3 x_5 - 300(-2+3r)x_2^2 x_5^2 \\ -60(-2+5r)x_2^4 x_6 \\ +360(-16+r)x_2 x_3^2 x_6 \\ +3600(-4+r)x_3 x_5 x_6 \\ +7200(2+r)x_2 x_6^2 \end{array} \right) \\ -\frac{r+5}{2^6 \cdot 3^4 \cdot 5^4} \left(\begin{array}{c} x_2^4 x_3 + 27x_2 x_3^3 - 20x_2^3 x_5 \\ -30x_3^2 x_5 - 60x_2 x_3 x_6 \\ +600x_5 x_6 \end{array} \right) V^3 \\ +\frac{r+5}{2^4 \cdot 3^3 \cdot 5^5}x_2(x_2^4 + 27x_2 x_3^2 - 120x_3 x_5 - 60x_2 x_6)V^4 \\ +\frac{3+2r}{1800}x_2 x_5 V^7 \end{array} \right] u,$$

$$(V^3 V^4 + V^4 V^3)u = \left[\begin{array}{l} \frac{2-r}{9000}x_2\{(20+3r)x_2 x_3 + 10(-4+3r)x_5\} \\ -\frac{7+2r}{360}x_2^2 V^3 + \frac{3+r}{25}x_3 V^4 + (3+2r)V^7 \end{array} \right] u,$$

$$(V^3V^7 + V^7V^3)u = \begin{bmatrix} \frac{r-2}{2^4 \cdot 3^3 \cdot 5^5} \begin{pmatrix} 4rx_2^5 + 45(5+r)x_2^2x_3^2 \\ -90(14+3r)x_2x_3x_5 \\ +1500x_5^2 - 60(-2+3r)x_2^2x_6 \end{pmatrix} \\ -\frac{(13+2r)}{1800}x_2x_5V^3 - \frac{(3+r)}{3750}(2x_2^3 - 9x_3^2 - 60x_6)V^4 \\ -\frac{(3+2r)}{50}x_3V^7 \end{bmatrix} u,$$

$$(V^4V^7 + V^7V^4)u = \begin{bmatrix} \frac{r-2}{2^6 \cdot 3^4 \cdot 5^4} \begin{pmatrix} (20+r)x_2^4x_3 - 36(5+r)x_2x_3^3 \\ +10(-4+5r)x_2^3x_5 - 60(2+r)x_3^2x_5 \\ +2160x_2x_3x_6 - 1200(2+r)x_5x_6 \end{pmatrix} \\ +\frac{(7+2r)}{43200}(x_2^4 + 6x_2x_3^2 - 40x_3x_5 - 40x_2x_6)V^3 \\ -\frac{(7+r)}{900}x_2x_5V^4 + \frac{(3+2r)}{360}x_2^2V^7 \end{bmatrix} u.$$

The characteristic exponents of the system $\mathcal{M}(r, s)$ along the hypersurface $f_0 = 0$ are

$$\frac{1}{30}(s-r+2), \frac{1}{30}(s-r+2), \frac{1}{30}(s-r+2), \frac{1}{30}(s+3r+8).$$

The system $\mathcal{M}(r, s)$ possibly has a non-trivial quotient $\mathcal{D}$-module. By direct computation we find the following:

(1) $\mathcal{M}(2, s)$ has a quotient defined by $V^0u = su, V^3u = V^4u = V^7u = 0$.

(2) $\mathcal{M}(-1, s)$ has a quotient added by $(V^3 - \frac{3}{50}x_3)u = 0$.

(3) $\mathcal{M}(-2, s)$ has a quotient added by $(V^4 + \frac{1}{90}x_2^2)u = 0$.

(4) $\mathcal{M}(-5, s)$ has a quotient added by $(V^7 + \frac{7}{1800}x_2x_5)u = 0$.

In the case (1), the quotient has a solution $u = f_0^{s/30}$. On the other hand, in the cases (2), (3), (4), there are three fundamental solutions outside $f_0(x_2, x_3, x_5, x_6) = 0$.

References

1. D. Bessis and J. Michel, Explicit presentations for exceptional braid groups. Experimental Math., **13** (2004), 257-266.

2. Y. Haraoka and M. Kato, Generating systems for finite irreducible complex reflection groups. Funkcialaj Ekvacioj **53** (2010), 435-488.

3. M. Kato and J. Sekiguchi, Systems of uniformization equations with respect to the discriminant sets of complex reflection groups of rank three. Preprint.

4. P. Orlik and H. Terao. *Arrangements of Hyperplanes.* Grundlehren der mathematisches Wissenschaften 300, Berlin: Springer-Verlage, 1992.

5. K. Saito, On the uniformization of complements of discriminant loci. RIMS Kokyuroku **287** (1977), 117-137.

6. K. Saito, Theory of logarithmic differential forms and logarithmic vector fields. J. Faculty of Sciences, Univ. Tokyo **27** (1980), 265-291.

7. K. Saito, Uniformization of orbifold of a finite reflection group. In "*Frobenius Manifold, Quantum Cohomology and Singularities.*" A Publication of the Max-Planck-Institute, Mathematics, Bonn, 265-320.

8. J. Sekiguchi, Three dimensional Saito free divisors and deformations of singular curves. J. Siberian Federal Univ., Mathematics and Physics, **1** (2008), 33-41.

9. J. Sekiguchi, A classification of weighted homogeneous Saito free divisors. J. Math. Soc. Japan, **61** (2009), 1071-1095.

10. J. Sekiguchi, Systems of uniformization equations along Saito free divisors and related topics, in *"The Third Japanese-Australian Workshop on Real and Complex Singularities"*, Proceedings of the Centre for Mathematics and its Applications, **43** (2010), 83-126.

11. J. Sekiguchi, Systems of uniformization equations and dihedral groups. Kumamoto J. Math., **23** (2010), 7-26.

12. J. Sekiguchi, Systems of uniformization equations and hyperelliptic integrals. J. Math. Sci., **175** (2011), 57-79.

13. J. Sekiguchi, The dicriminant of the reflection group of type D_4 and holonomic systems with singularities along its zero locus. To appear in Tohoku Math. J.

14. G. C. Shephard and A. J. Todd, Finite reflection groups. Canad. J. Math., **6** (1954), 274-304.

15. T. Yano and J. Sekiguchi, The microlocal structure of weighted homogeneous polynomials associated with Coxeter systems. I, II, Tokyo J. Math. **2** (1979), 193-219, Tokyo J. Math. **4** (1981), 1-34.

Holonomic systems of differential equations of rank two with singularities along Saito free divisors of simple type

Jiro Sekiguchi*

Department of Mathematics, Tokyo University of Agriculture and Technology, Tokyo 184-8588, Japan

`sekiguti@cc.tuat.ac.jp`

We will show many examples of holonomic systems of three variables of rank two with singularities along Saito free divisors of simple type. Solutions of some of them are expressed in terms of elementary functions or hypergeometric functions. But some of the holonomic systems are difficult to solve. In the last section, we discuss the relationship between the holonomic systems obtained in this paper and algebraic solutions of Painlevé sixth equation.

Keywords: Saito free divisor, holonomic system
AMS classification numbers: 34M56, 20F65

1. Introduction

The notion of Saito free divisors is introduced by K. Saito around 1970's. One of the reasons why he introduced it is to describe characteristic properties of the parameter space of versal deformations of isolated singularities. He also formulated the notion of systems of uniformization equations with logarithmic poles along Saito free divisors. These were published in a short note [4] (see also [5], [6]). After his pioneering work, a lot of studies related with Saito free divisors were done in various branches of mathematics; hyperplane arrangements, topologies of complements of Saito free divisors, flat structure or Frobenius manifold structure, etc.

As to systems of uniformization equations, the existence of such a system depends on the choice of a Saito free divisor. It is A. G. Aleksandrov who pointed out this basic result (cf. [1]) and such systems were studied in [9], [3]. In this paper, we will study holonomic systems of differential equations on $\mathbf{C}^3$ of rank two which have singularities along Saito free divisors. These

---

*Partially supported by Grand-in-Aid for Scientific Research (No. 2354077), Japan Society of the Promotion of Science.

systems are rank two analogue of systems of uniformization equations but it is easier to treat compared with uniformization equations.

We start the explanation of this paper with defining Saito free divisors. Let $\mathbf{C}^n$ be an n-dimensional affine space with coordinate system $x = (x_1, x_2, \ldots, x_n)$ and let $R = \mathbf{C}[x_1, x_2, \ldots, x_n]$ be its coordinate ring. Let $F(x) = F(x_1, x_2, \ldots, x_n)$ be a reduced polynomial. Then the hypersurface $F(x) = 0$ is Saito free if the following condition (C.1) is satisfied:

(C.1) *There are vector fields*

$$V_i = \sum_{j=1}^{n} a_{ij}(x) \partial_{x_j} \ (i = 1, 2, \ldots, n)$$

such that $V_i F / F \in R \ (i = 1, 2, \ldots, n)$ *and* $\det M = cF$ *for a non-zero constant* c, *where* M *is an* $n \times n$ *matrix whose* (i, j)-*entry is* $a_{ij}(x)$. *The matrix* M *is called a generating matrix.*

We now assume that $F(x)$ is weighted homogeneous, namely, there is a vector field $E = \sum_{j=1}^{n} d_j x_j \partial_{x_j}$ such that $EF = dF$, where $d_1, d_2, \ldots, d_n, d$ are positive integers. Moreover we assume that $0 < d_1 < d_2 < \ldots < d_n$. Under these assumptions, we may take $V_i \ (i = 1, 2, \ldots, n)$ so that $V_1 = E$ and $[E, V_i] = k_i V_i$ with constants $k_i \ (i = 1, 2, \ldots, n)$ such that $k_1 = 0 \leq k_2 \leq \cdots \leq k_n$.

We next introduce the notion of systems of uniformization equations. Let $F(x)$ be Saito free and let $D : F(x) = 0$ be the hypersurface defined by $F(x)$. Define

$$\begin{cases} V_1 u = su, \\ V_i V_j u = \left(\sum_{k=2}^{n} p_{ij}^k(x) V_k + q_{ij}(x) \right) u & (i, j = 2, 3, \ldots, n), \end{cases} \tag{1.1}$$

where $p_{ij}^k(x), q_{ij}(x) \in R$, $s \in \mathbf{C}$. If (1.1) is integrable, it is called a system of uniformization equations. The system (1.1) has singularities along D.

We restrict our attention to the case $n = 3$. Let $F(x, y, z)$ be one of the seventeen polynomials given in [8], Theorem 1 and let M be the matrix given in [8], Appendix. In the following we write x_1, x_2, x_3 in place of x, y, z. We define vector fields $V_j \ (j = 1, 2, 3)$ by $^t(V_1, V_2, V_3) = M^t(\partial_{x_1}, \partial_{x_2}, \partial_{x_3})$.

Let $u(x_1, x_2, x_3)$ be an unknown function and define two kinds of systems of differential equations. The first one is

$$\begin{cases} V_1 u = su, \\ V_2 u = p_1(x)u, \\ V_3 V_3 u = (p_2(x) + p_3(x)V_3)u, \end{cases} \tag{1.2}$$

and the second one is

$$\begin{cases} V_1 u & = su, \\ V_2 V_2 u = (p_1(x) + p_2(x)V_2)u, \\ V_3 u & = (p_3(x) + p_4(x)V_2)u. \end{cases} \tag{1.3}$$

The former is called of Type I and the latter is called of Type II. In both cases, s is a complex number and $p_1(x)$, $p_2(x), \ldots$ are polynomials of $x = (x_1, x_2, x_3)$. The purpose of this paper is to show the result on a classification of holonomic systems of Type I and Type II for Saito free divisors introduced in [8]. Since the Saito free divisors in [8] are regarded as 1-parameter deformations of simple curve singularities of exceptional type, they are called Saito free divisors of simple type in this paper. A basic question is whether the holonomic systems obtained in this paper are reduced to elementary ones or not. We also discuss on this question in the main text. Solutions of some of holonomic systems thus obtained are expressed by the products of complex powers of polynomials and their integrals. This kind of phenomenon reflects the solvability of groups realized as two dimensional representations of fundamental groups of complements of Saito free divisors. The fundamental groups in these cases are studied in [7]. Let $D : F = 0$ be a Saito free divisor and let $\mathcal{M}$ be one of holonomic systems treated in this paper. We note that there are two multi-valued holomorphic fundamental solutions u_1, u_2 of $\mathcal{M}$ on $\mathbf{C}^3 - D$. Then the second basic question is to determine the image of the so called Schwarz map defined by $\varphi : \mathbf{C}^3 - D \ni x = (x_1, x_2, x_3) \to (u_1(x) : u_2(x)) \in \mathbf{P}^1$. As an attempt to answer to this question we study the restriction of u_1, u_2 to $x_1 = 0$ and $x_2 = 0$. In some cases the restrictions are expressed by hypergeometric functions.

We explain the contents of this paper briefly. In section 2, we introduce the notion of holonomic systems of Type I and those of Type II for Saito free divisors in $\mathbf{C}^3$ and their integrability conditions. In section 3, we will construct holonomic systems of Type I/Type II for Saito free divisors defined as the zero sets of seventeen polynomials in [8]. In section 4, we discuss the meaning of the systems constructed in section 3 and in particular show a relationship between the holonomic systems treated in this paper and algebraic solutions of Painlevé VI equation taking as an example the case of the holonomic system in subsection 3.9.

In closing this introduction, we note that our computations in this paper are performed using the computer algebraic software Mathematica.

Acknowledgement: The author thanks the referee for reading the contents carefully. In particular, being pointed out by him, the author recognized that there are many mistakes of the computations on the systems of Types I and II in the first draft and could correct them.

2. Holonomic systems of rank two

Let $R = \mathbf{C}[x_1, x_2, x_3]$ be a polynomial ring of three variables and let $M = (m_{ij}(x_1, x_2, x_3))_{i,j=1,2,3}$ be a 3×3 matrix whose matrix entries are polynomials. Let V_1, V_2, V_3 be vector fields defined by M as in section 1, namely,

$$^t(V_1, V_2, V_3) = M^t(\partial_{x_1}, \partial_{x_2}, \partial_{x_3}).$$

We assume that $m_{1j} = d_j x_j$ $(j = 1, 2, 3)$, where d_1, d_2, d_3 are integers such that $0 < d_1 < d_2 < d_3$. A polynomial $g(x) \in R$ is weighted homogeneous of degree k if $V_1 g = kg$. In this case we write $\deg g = k$. We assume that each m_{ij} is weighted homogeneous. A vector field V with polynomial coefficients is weighted homogeneous of degree d if $[V_1, V] = dV$. In this case, we say d the degree of V and write $\deg V = d$. In the sequel, we assume that V_1, V_2, V_3 are weighted homogeneous and that $\deg V_1 = 0 < \deg V_2 < \deg V_3$.

As is explained in the introduction, the polynomial $F(x_1, x_2, x_3) = \det M$ plays a central role in the subsequent argument. We assume that $F(x)$ is reduced and that V_1, V_2, V_3 are logarithmic along the hypersurface $D : F(x) = 0$, that is, $V_j F/F \in R$ $(j = 1, 2, 3)$. As an easy consequence, there are polynomials $c_j(x) \in R$ such that $V_j F = c_j F$ $(j = 1, 2, 3)$. In particular c_1 is a positive integer.

Under the above assumption, it follows from [5] that $\mathcal{L}_M = RV_1 + RV_2 + RV_3$ is a Lie algebra over R. In particular,

$$\begin{aligned}
[V_1, V_i] &= (\deg V_i)V_i \quad (i = 2, 3), \\
[V_2, V_3] &= w_1(x)V_1 + w_2(x)V_2 + w_3(x)V_3,
\end{aligned} \tag{2.4}$$

where w_j are weighted homogeneous polynomials.

We now introduce two kinds of systems of differential equations. Let $u = u(x_1, x_2, x_3)$ be an unknown function such that $V_1 u = su$ for a complex number s.

To define the first one, we consider the R-module $\mathcal{N} = Ru + RV_3 u$. We assume that u and $V_3 u$ are linearly independent over R and that $\mathcal{L}_M \mathcal{N} \subset \mathcal{N}$. Then u is a solution of the system of differential equations

$$\begin{cases}
V_1 u &= su, \\
V_2 u &= p_1(x)u, \\
V_3 V_3 u &= (p_2(x) + p_3(x)V_3)u,
\end{cases} \tag{2.5}$$

where $p_j(x) \in R$. The system (2.5) is called of Type I.

We take a solution u of (2.5) and put $\boldsymbol{u}_I = \begin{pmatrix} u \\ V_3 u \end{pmatrix}$. We rewrite (2.5) to a system for $\boldsymbol{u}_I$. Since $[V_1, V_3] = (\deg V_3)V_3$ and since $V_1 u = su$, we have

$$V_1 \boldsymbol{u}_I = \begin{pmatrix} s & 0 \\ 0 & s + \deg V_3 \end{pmatrix} \boldsymbol{u}_I.$$

Our next job is to compute $V_2 \boldsymbol{u}_I$. Since

$$V_3 V_2 u = V_3(p_1 u) = p_1 V_3 u + [V_3, p_1]u,$$

it follows that

$$V_2 V_3 u = [V_2, V_3]u + V_3 V_2 u = (w_1 V_1 + w_2 V_2 + w_3 V_3)u + (p_1 V_3 + [V_3, p_1])u.$$

Then

$$V_2 \boldsymbol{u}_I = \begin{pmatrix} V_2 u \\ V_2 V_3 u \end{pmatrix} = \begin{pmatrix} p_1 & 0 \\ sw_1 + p_1 w_2 + [V_3, p_1] & w_3 + p_1 \end{pmatrix} \boldsymbol{u}_I.$$

Our last job is to compute $V_3 \boldsymbol{u}_I$. In this case,

$$V_3 \boldsymbol{u}_I = \begin{pmatrix} V_3 u \\ V_3 V_3 u \end{pmatrix} = \begin{pmatrix} V_3 u \\ (p_2 + p_3 V_3)u \end{pmatrix} = \begin{pmatrix} 0 & 1 \\ p_2 & p_3 \end{pmatrix} \boldsymbol{u}_I.$$

As a consequence, we have

$$\mathcal{M}_I : V_i \boldsymbol{u}_I = A_i \boldsymbol{u}_I \quad (i = 1, 2, 3), \tag{2.6}$$

where A_1, A_2, A_3 are 2×2 matrices of the forms

$$A_1 = \begin{pmatrix} s & 0 \\ 0 & s + \deg V_3 \end{pmatrix},$$

$$A_2 = \begin{pmatrix} p_1 & 0 \\ sw_1 + p_1 w_2 + [V_3, p_1] & w_3 + p_1 \end{pmatrix},$$

$$A_3 = \begin{pmatrix} 0 & 1 \\ p_2 & p_3 \end{pmatrix}.$$

To define the second one, we consider the R-module $\mathcal{N} = Ru + RV_2 u$. We assume that u and $V_2 u$ are linearly independent over R and that $\mathcal{L}_M \mathcal{N} \subset \mathcal{N}$. Then u is a solution of the system of differential equations

$$\begin{cases} V_1 u &= su, \\ V_2 V_2 u &= (p_1(x) + p_2(x)V_2)u, \\ V_3 u &= (p_3(x) + p_4(x)V_2)u, \end{cases} \tag{2.7}$$

where $p_j(x) \in R$. The system (2.7) is called of Type II. We take a solution u of (2.7) and put $\boldsymbol{u}_{II} = \begin{pmatrix} u \\ V_2 u \end{pmatrix}$. We can compute $V_i \boldsymbol{u}_{II}$ $(i = 1, 2, 3)$ by an argument similar to the argument above. Then from (2.7), we find the system for $\boldsymbol{u}_{II}$ to be

$$\mathcal{M}_{II} : V_i \boldsymbol{u}_{II} = A_i \boldsymbol{u}_{II} \quad (i = 1, 2, 3) \tag{2.8}$$

where

$$A_1 = \begin{pmatrix} s & 0 \\ 0 & s + \deg V_2 \end{pmatrix},$$

$$A_2 = \begin{pmatrix} 0 & 1 \\ p_1 & p_2 \end{pmatrix},$$

$$A_3 = \begin{pmatrix} p_3 & p_4 \\ -sw_1 + [V_2, p_3] + p_1 p_4 - p_3 w_3 & -w_2 - p_4 w_3 + p_3 + p_2 p_4 + [V_2, p_4] \end{pmatrix}.$$

Both of $\mathcal{M}_I$ and $\mathcal{M}_{II}$ are written in the form

$$\mathcal{M} : V_i \boldsymbol{u} = A_i \boldsymbol{u} \quad (i = 1, 2, 3) \tag{2.9}$$

where A_1, A_2, A_3 are 2×2 matrices. With the help of the relations (2.4), we obtain the following compatibility conditions on the matrices A_1, A_2, A_3:

$$\begin{aligned} {[A_1, A_2]} + [V_1, A_2] - (\deg V_2)A_2 &= O, \\ [A_1, A_3] + [V_1, A_3] - (\deg V_3)A_3 &= O, \\ [A_2, A_3] - [V_2, A_3] + [V_3, A_2] + w_1(x)A_1 + w_2(x)A_2 + w_3(x)A_3 &= O, \end{aligned} \tag{2.10}$$

where we note that A_1 is a constant matrix. The first and the second relations of (2.10) imply that each matrix entry of A_2, A_3 are weighted homogeneous. On the other hand, the third one implies non-trivial equations among the matrix entries of A_2, A_3.

The relations (2.10) are rewritten in a familiar form which we are going to explain. Since $^t(V_1, V_2, V_3) = M^t(\partial_{x_1}, \partial_{x_2}, \partial_{x_3})$, we introduce 2×2 matrices B_1, B_2, B_3 by

$$^t(B_1, B_2, B_3) = M^{-1t}(A_1, A_2, A_3). \tag{2.11}$$

Then FB_j has polynomial entries and the system (2.9) is equivalent to

$$\partial_{x_j} \boldsymbol{u} = B_j \boldsymbol{u} \quad (j = 1, 2, 3). \tag{2.12}$$

The integrability condition for (2.12) is

$$[B_i, B_j] + \frac{\partial B_i}{\partial x_j} - \frac{\partial B_j}{\partial x_i} = O \quad (\forall i, j). \tag{2.13}$$

It is clear that (2.13) is also an integrability condition for $\mathcal{M}$.

Remark 2.1. Let u be an unknown function such that $V_1 u = su$ for a complex number s. Replacing the R-module $\mathcal{N}$ with different one, many examples of systems of differential equations are produced. In this remark, we introduce two such R-modules which are interesting to study.

As the first one, we take the R-module $\mathcal{N} = Ru$. If $\mathcal{L}_M \mathcal{N} \subset \mathcal{N}$, then u satisfies the system of equations

$$\begin{cases} V_1 u = su, \\ V_2 u = p_1(x)u, \\ V_3 u = p_2(x)u, \end{cases} \tag{2.14}$$

where $p_j(x) \in R$ $(j = 1, 2)$. On the other hand, we have

$$V_2 V_3 u = V_2(p_2 u) = p_2 V_2 u + [V_2, p_2]u = p_2 p_1 u + [V_2, p_2]u,$$
$$V_3 V_2 u = V_3(p_1 u) = p_1 V_3 u + [V_3, p_1]u = p_1 p_2 u + [V_3, p_1]u.$$

Then

$$[V_2, V_3]u = (w_2 V_2 + w_3 V_3)u = (w_2 p_1 + w_3 p_2)u$$

and as a consequence,

$$[V_2, p_2] - [V_3, p_1] = w_2 p_1 + w_3 p_2$$

is the integrability condition of (2.14). The most interesting example of the system (2.14) is the system which has the solution $u = F^{s/\deg F}$. There are a lot of studies on $\mathcal{D}$-modules which govern complex powers of polynomials.

As the second one, we take $\mathcal{N} = Ru + RV_2 u + RV_3 u$, keeping the assumption $V_1 u = su$. If $\mathcal{L}_M \mathcal{N} \subset \mathcal{N}$, it is easy to see that u satisfies a system of uniformization equations (cf. (1.1)).

In this manner, it is easy to define various kinds of systems of differential equations related with Saito free divisors. But it is not so easy to find Saito free divisors such that there exist non-trivial systems of differential equations related with R-modules defined in this section. Our main purpose of this paper is to construct systems of Type I and/or Type II with respect to Saito free divisors of simple type. Moreover we study solutions of the systems. Some of solutions of the systems are expressed in terms of elementary functions or hypergeometric functions. But some seem not to be done by known functions.

We now explain how to obtain the systems of Type I/II by taking the case of the polynomial $F_{H,1}$ of §3.8 below as an example.

We discuss here only the case of systems of Type I. Our effort is focused on the systems $\mathcal{M}_I$ of (2.6), because the system (2.5) is recovered from $\mathcal{M}_I$. In this case, variables x_1, x_2, x_3 have weights 1, 3, 5, respectively.

To determine $\mathcal{M}_I$, we first introduce polynomials φ_j ($j = 1, 2, \ldots, 5$) of x_1, x_2, x_3 such that

$$A_1 = \begin{pmatrix} s & 0 \\ 0 & s+4 \end{pmatrix},$$

$$A_2 = \begin{pmatrix} \varphi_1 & 0 \\ \varphi_2 & \varphi_3 \end{pmatrix},$$

$$A_3 = \begin{pmatrix} 0 & 1 \\ \varphi_4 & \varphi_5 \end{pmatrix},$$

where $s \in \mathbf{C}$. By the weight condition, it is easy to see that $\varphi_1, \varphi_2, \varphi_3, \varphi_4, \varphi_5$ are weighted homogeneous of degree 2,6,2,8,4, respectively. Then we may take

$$\varphi_1 = c_1 x_1^2,$$
$$\varphi_2 = c_2 x_1 x_3 + c_3 x_2^2 + c_4 x_1^3 x_2 + c_5 x_1^6,$$
$$\varphi_3 = c_6 x_1^2,$$
$$\varphi_4 = c_7 x_2 x_3 + c_8 x_1^3 x_3 + c_9 x_1^2 x_2^2 + c_{10} x_1^5 x_2 + c_{11} x_1^8,$$
$$\varphi_5 = c_{12} x_1 x_2 + c_{13} x_1^4,$$

where $c_1, c_2, \ldots, c_{13}$ are constants. Next, we define B_1, B_2, B_3 from A_1, A_2, A_3 by the equation (2.11). Then the integrability condition (2.13) is equivalent to a system of algebraic equations among $s, c_1, c_2, \ldots, c_{13}$. Solving this system, we obtain the polynomials φ_j ($j = 1, 2, 3, 4, 5$) for the system $\mathcal{M}_I$ of (2.6) to be integrable.

3. Construction of holonomic systems of rank two

The Saito free divisors treated in this paper are seventeen divisors defined as hypersurfaces constructed by the author [8] where he classified weighted homogeneous polynomials of three variables under certain conditions. They are named $F_{A,1}$, $F_{A,2}$, $F_{B,j}$ ($j = 1, 2, \ldots, 7$), $F_{H,j}$ ($j = 1, 2, \ldots, 8$) and are closely related with irreducible real reflection groups of rank three. Their types are A_3, B_3, H_3. In fact $F_{X,1}$ is regarded as the discriminant of the reflection group of type X_3 ($X = A, B, H$). Moreover they are also related with simple curve singularities of types E_6, E_7, E_8. In particular each of the hypersurfaces in [8] is regarded as a 1-parameter family of simple curve singularity whose type is one of E_6, E_7, E_8. For this reason, the divisors treated in this paper are called *Saito free divisors of simple type*.

The purpose of this section is to construct systems of Type I/Type II with singularities along Saito free divisors which are defined by $F_{X,j}$ ($X = A, B, H$). Throughout this section, for a given polynomial $F_{X,j}$ defining a

Saito free divisor, M is a generating matrix of $F_{X,j}$ and V_1, V_2, V_3 are vector fields defined by M. Note that there are many choices of the matrix M and you find in [8] a standard one of the Saito matrices for each polynomial $F_{X,j}$. Since it is possible to show that if $F_{X,j}$ is equal to one of $F_{B,j}$ ($j = 5, 7$), $F_{H,j}$ ($j = 4, 5, 6, 7, 8$), there is no holonomic system neither of Type I nor of Type II, we only treat the polynomials $F_{A,j}$ ($j = 1, 2$), $F_{B,j}$ ($j = 1, 2, 3, 4, 6$), $F_{H,j}$ ($j = 1, 2, 3$).

3.1. The case $F_{A,1}$

The polynomial is given by

$$F_{A,1} = -4x_1^3 x_2^2 - 27x_2^4 + 16x_1^4 x_3 + 144x_1 x_2^2 x_3 - 128x_1^2 x_3^2 + 256x_3^3$$

and its generating matrix is

$$M = \begin{pmatrix} 2x_1 & 3x_2 & 4x_3 \\ 3x_2 & -x_1^2 + 4x_3 & -\frac{1}{2}x_1 x_2 \\ 4x_3 & -\frac{1}{2}x_1 x_2 & \frac{1}{4}(8x_1 x_3 - 3x_2^2) \end{pmatrix}.$$

Note that $F_{A,1}$ is regarded as the discriminant of the reflection group of type A_3.

There are a holonomic system of Type I and a holonomic system of Type II.

<u>System of Type I</u>

$$\begin{cases} V_1 u = r_1 u, \\ V_2 u = 0, \\ V_3 V_3 u = \frac{1}{36}\{-(r_1 + r_2 + 1)(r_1 - r_2 + 1)x_1^2 + 12(2r_1 + r_2^2 - 1)x_3\}u \\ \quad + \frac{1}{3}(1 + r_1)x_1 V_3 u. \end{cases}$$

$$(3.15)$$

In this case, solutions of the system (3.15) are expressed in terms of hypergeometric functions. We explain this result obtained by K. Saito (cf. [4]) briefly. Put

$$P = 4x_3 + \frac{1}{3}x_1^2, \quad Q = \frac{1}{2}(x_2^2 - \frac{8}{3}x_1 x_3 + \frac{2}{27}x_1^3).$$

Then

$$\det M = -P^3 + 27Q^2.$$

Let V_1, V_2, V_3 be vector fields defined by M. Putting $V_3' = V_3 - \frac{1}{6}x_1 V_1$, we have

$$V_3' P = -6Q, \quad V_3' Q = -\frac{1}{3}P^2.$$

The system (3.15) turns out to be

$$\begin{cases} V_1 u & = r_1 u, \\ V_2 u & = 0, \\ V_3' V_3' u & = \frac{r_2^2 - 1}{12} P u \end{cases}$$

and

$$u = (P^3 - 27Q^2)^{r_1/12} F\left(\frac{1+r_2}{12}, \frac{1-r_2}{12}, \frac{2}{3}; \frac{P^3}{P^3 - 27Q^2}\right)$$

is one of its solutions, where $F(a, b, c; x)$ is a hypergeometric function.

System of Type II

$$\begin{cases} V_1 u & = r_1 u, \\ V_3 u & = \frac{1}{6}(r_1 - 1)x_1 u, \\ V_2 V_2 u & = -\frac{1}{2}x_1 u. \end{cases} \tag{3.16}$$

In spite that it is not clear whether there is a solution of the system (3.16) which is expressed in terms of elementary functions or special functions, it can be shown that the restriction of any solution of the system (3.16) to the hypersurface $x_1 = 0$ is expressed by hypergeometric functions. In fact, there is a solution of (3.16) such that

$$u|_{x_1=0} = x_3^{\frac{1}{3}}(-27x_2^4 + 256x_3^3)^{\frac{1}{12}(r_1-1)} F\left(-\frac{1}{12}, \frac{7}{12}, \frac{3}{4}; \frac{27x_2^4}{256x_3^3}\right).$$

On the other hand, we obtain differential equations

$$\begin{cases} \partial_{x_1}\begin{pmatrix} u \\ V_2 u \end{pmatrix} = \frac{1}{6(x_1^2 - 4x_3)}\begin{pmatrix} (2r_1 + 1)x_1 & 0 \\ 0 & (2r_1+1)x_1 \end{pmatrix}\begin{pmatrix} u \\ V_2 u \end{pmatrix} \\ \partial_{x_3}\begin{pmatrix} u \\ V_2 u \end{pmatrix} = \frac{1}{12x_3(x_1^2 - 4x_3)}\begin{pmatrix} (r_1-1)x_1^2 & 0 \\ -12r_1x_3 & \\ & (r_1+2)x_1^2 \\ 0 & -12(r_1+1)x_3 \end{pmatrix}\begin{pmatrix} u \\ V_2 u \end{pmatrix} \end{cases} \tag{3.17}$$

by restricting the system (3.16) to $x_2 = 0$. As a consequence, we obtain a solution

$$u|_{x_2=0} = c_1 x_3^{\frac{1}{12}(r_1-1)}(4x_3 - x_1^2)^{\frac{1}{12}(2r_1+1)},$$

$$(V_2 u)|_{x_2=0} = c_2 x_3^{\frac{1}{12}(r_1+2)}(4x_3 - x_1^2)^{\frac{1}{12}(2r_1+1)}$$

for some constants c_1, c_2 of the differential equations (3.17).

3.2. *The case $F_{A,2}$*

The polynomial is

$$F_{A,2} = 2x_1^6 + 18x_1^3x_2^2 + 27x_2^4 - 3x_1^4x_3 - 18x_1x_2^2x_3 + x_3^3$$

and its generating matrix is

$$M = \begin{pmatrix} 2x_1 & 3x_2 & 4x_3 \\ 3x_2 & \frac{1}{2}(x_3 - x_1^2) & 6x_1x_2 \\ 4x_3 & -2x_1x_2 & 16x_1^3 + 24x_2^2 - 8x_1x_3 \end{pmatrix}.$$

There is a holonomic system of Type I but no holonomic system of Type II.

System of Type I

$$\begin{cases} V_1 u &= r_1 u, \\ V_2 u &= 0, \\ V_3 V_3 u &= \frac{4}{3}\{-(3r_1^2 + 6r_1 - r_2^2 + 1)x_1^2 - (6r_1 + r_2^2 - 1)x_3\}u \\ & \quad -4(r_1 + 1)x_1 V_3 u. \end{cases} \tag{3.18}$$

In this case, it is easy to solve the system (3.18) by an argument parallel to that for the case $F_{A,1}$. We explain it briefly. First put

$$P = x_1^2 - x_3, \quad Q = \frac{1}{3}(3x_2^2 - x_1x_3 + x_1^3).$$

Then $\det M$ coincides with $P^3 - 27Q^2$ up to a constant factor. Let V_1, V_2, V_3 be vector fields defined by M. Putting $V_3' = V_3 + 2x_1V_1$, we have

$$V_3'P = -24Q, \quad V_3'Q = -\frac{4}{3}P^2.$$

The system (3.18) turns out to be

$$\begin{cases} V_1 u &= r_1 u, \\ V_2 u &= 0, \\ V_3'V_3'u &= \frac{4}{3}(r_2^2 - 1)Pu, \end{cases}$$

and

$$u = (P^3 - 27Q^2)^{r_1/12}F\left(\frac{1+r_2}{12}, \frac{1-r_2}{12}, \frac{2}{3}; \frac{P^3}{P^3 - 27Q^2}\right)$$

is one of its solutions.

3.3. The case $F_{B,1}$

The polynomial is

$$F_{B,1} = x_3(-x_1^2 x_2^2 + 4x_2^3 + 4x_1^3 x_3 - 18x_1 x_2 x_3 + 27x_3^2)$$

and its generating matrix is

$$M = \begin{pmatrix} x_1 & 2x_2 & 3x_3 \\ 2x_2 & x_1 x_2 + 3x_3 & 2x_1 x_3 \\ 3x_3 & 2x_1 x_3 & x_2 x_3 \end{pmatrix}.$$

Note that $F_{B,1}$ is regarded as the discriminant of the reflection group of type B_3.

There are three holonomic systems of rank two. One is of Type I and the remaining two are of Type II.

System of Type I

$$\begin{cases} V_1 u & = r_1 u, \\ V_2 u & = \frac{1}{6}(2r_1 + 6r_2 - 1)x_1 u, \\ V_3 V_3 u & = \frac{1}{4}\{-4(r_2 - r_3)(r_2 + r_3)x_2^2 + (8r_2 - 12r_3^2 - 1)x_1 x_3\}u \\ & \quad + 2r_2 x_2 V_3 u. \end{cases} \tag{3.19}$$

In this case, the restrictions to $x_1 = 0$ and $x_2 = 0$ of solutions of (3.19) are expressed in terms of hypergeometric functions. We first treat the case $x_1 = 0$. There is a solution u of (3.19) such that

$$u|_{x_1=0} = (4x_2^3 + 27x_3^2)^{\frac{1}{6}(1+r_1-3r_2+3r_3)} x_3^{r_2-r_3-\frac{5}{3}} x_2^2$$

$$\times F\left(\frac{5}{6}, r_3 + \frac{5}{6}; \frac{5}{3}; -\frac{4x_2^3}{27x_3^2}\right).$$

We next treat the case $x_2 = 0$. There is a solution u of (3.19) such that

$$u|_{x_2=0} = (4x_1^3 + 27x_3)^{\frac{1}{6}(1+r_1-3r_2+3r_3)} x_3^{\frac{1}{6}(r_1+3r_2-3r_3-1)}$$

$$\times F\left(\frac{6r_3+1}{12}, \frac{6r_3+7}{12}; \frac{2}{3}; -\frac{4x_1^3}{27x_3}\right).$$

Systems of Type II

There are two holonomic systems of Type II.

Type (II.1)

$$\begin{cases} V_1 u & = r_1 u, \\ V_2 V_2 u = \frac{1}{4}\{(r_2 - r_3)(r_2 + r_3)x_1^2 + (-1 + 4r_3 - 3r_2^2)x_2\}u \\ \qquad\qquad + r_3 x_1 V_2 u, \\ V_3 u & = -\frac{1}{6}(1 + 2r_1 - 3r_3)x_2 u. \end{cases} \qquad (3.20)$$

In this case, solutions of the system (3.20) are expressed in terms of hypergeometric functions. We will explain the result briefly. We first construct 2×2 matrices A_j $(j = 1, 2, 3)$ from (3.20) as we explained the way similar to do (2.8) from (2.7). Then we obtain a system

$$V_j \boldsymbol{u} = A_j \boldsymbol{u} \quad (j = 1, 2, 3), \qquad (3.21)$$

where $\boldsymbol{u} = \begin{pmatrix} u \\ V_2 u \end{pmatrix}$. We next define 2×2 matrices B_j $(j = 1, 2, 3)$ by the equation similar to (2.11). Then we obtain a system

$$\partial_{x_j} \boldsymbol{u} = B_j \boldsymbol{u} \quad (j = 1, 2, 3). \qquad (3.22)$$

We thirdly change the variables:

$$x_1 = y_1, \quad x_2 = \frac{1}{3}(3y_2 + y_1^2), \quad x_3 = \frac{1}{27}(27y_3 + 9y_1 y_2 + y_1^3).$$

Then (3.22) turns out to be

$$\partial_{y_j} \boldsymbol{u} = C_j \boldsymbol{u} \quad (j = 1, 2, 3), \qquad (3.23)$$

where C_j $(j = 1, 2, 3)$ are 2×2 matrices defined by using B_j $(j = 1, 2, 3)$. In particular $C_2 = B_2 + \frac{1}{3}x_1 B_3$, $C_3 = B_3$. Putting

$$v = x_3^{-\frac{1}{6}(-2r_1 + 3r_3 - 1)}(27y_3^2 + 4y_2^3)^{-\frac{1}{12}(2 + 4r_1 - 3r_2 - 3r_3)}u,$$

we find that v is a solution of

$$\left(\partial_{y_2}^2 - \frac{6(r_2 - 2)y_2^2}{4y_2^3 + 27y_3^2}\partial_{y_2} + \frac{(3r_2 - 1)(3r_2 - 7)y_2}{4(4y_2^3 + 27y_3^2)} \right)v = 0$$

from $\partial_{y_2}\boldsymbol{u} = C_2 \boldsymbol{u}$ and is also a solution of

$$\left(\partial_{y_3}^2 - \frac{27(r_2 - 2)y_3}{4y_2^3 + 27y_3^2}\partial_{y_3} + \frac{3(3r_2 - 1)(3r_2 - 5)}{4(4y_2^3 + 27y_3^2)} \right)v = 0$$

from $\partial_{y_3}\boldsymbol{u} = C_3 \boldsymbol{u}$. As a consequence, we obtain a solution

$$v = y_2^{\frac{1}{4}(3r_2 - 1)} F\left(\frac{1 - 3r_2}{12}, \frac{5 - 3r_2}{12}, \frac{1}{2}; -\frac{27y_3^2}{4y_2^3} \right)$$

of the two differential equations. Then

$$u = x_3^{\frac{1}{6}(-2r_1+3r_3-1)} f_0^{\frac{1}{12}(2+4r_1-3r_2-3r_3)} (3x_2 - x_1^2)^{\frac{1}{4}(3r_2-1)}$$

$$\times F\left(\frac{1-3r_2}{12}, \frac{5-3r_2}{12}, \frac{1}{2}; \frac{(27x_3 - 9x_1x_2 + 2x_1^3)^2}{4(x_1^2 - 3x_2)^3}\right)$$

is a solution of the system (3.20), where

$$f_0 = (-x_1^2 x_2^2 + 4x_2^3 + 4x_1^3 x_3 - 18x_1x_2x_3 + 27x_3^2).$$

Type (II.2)

$$\begin{cases} V_1u & = r_1 u, \\ V_2 V_2 u = -\frac{1}{4}r_2\{(2+r_2)x_1^2 - 4x_2\}u + (1+r_2)x_1 V_2 u, \\ V_3 u & = \frac{1}{6}\{-3r_2r_3x_1^2 + (2-2r_1+3r_2-6r_3)x_2\}u + r_3 x_1 V_2 u. \end{cases} \quad (3.24)$$

In this case, the restrictions to $x_1 = 0$ and $x_2 = 0$ of solutions of (3.24) are expressed in terms of hypergeometric functions. We first treat the case $x_1 = 0$. There is a solution u of (3.24) such that

$$u|_{x_1=0} = (4x_2^3 + 27x_3^2)^{\frac{1}{12}(2+4r_1-3r_2-6r_3)} x_3^{\frac{1}{6}(-2-2r_1+3r_2+6r_3)}$$

$$\times F\left(\frac{5}{6}, -r_3 + \frac{1}{3}, \frac{2}{3}; -\frac{4x_2^3}{27x_3^2}\right).$$

We next treat the case $x_2 = 0$. There is a solution u of (3.24) such that

$$u|_{x_2=0} = (4x_1^3 + 27x_3)^{\frac{1}{12}(2+4r_1-3r_2-6r_3)} x_3^{\frac{1}{12}(-2+3r_2+6r_3)}$$

$$\times F\left(\frac{-3r_3+1}{6}, \frac{-3r_3+4}{6}, \frac{1}{3}; -\frac{4x_1^3}{27x_3}\right).$$

3.4. The case $F_{B,2}$

The polynomial is

$$F_{B,2} = x_3(2x_2^3 - 4x_1^3 x_3 - 18x_1x_2x_3 - 27x_3^2)$$

and its generating matrix is

$$M = \begin{pmatrix} x_1 & 2x_2 & 3x_3 \\ 2x_2 & -\frac{2}{3}(2x_1x_2 - 9x_3) & -4x_1x_3 \\ 3x_3 & -\frac{2}{3}(x_2^2 + 3x_1x_3) & -2x_2x_3 \end{pmatrix}.$$

There are four holonomic systems of rank two. One is of Type I and the remaining three are of Type II. In this subsection, we always put

$$f_0 = 2x_2^3 - 4x_1^3 x_3 - 18x_1x_2x_3 - 27x_3^2.$$

System of Type I

$$\begin{cases} V_1 u = r_1 u, \\ V_2 u = r_2 x_1 u, \\ V_3 V_3 u = -\frac{r_2}{12}\{(-4+3r_2)x_2^2 + 12x_1 x_3\}u + \frac{1}{3}(-4+3r_2)x_2 V_3 u. \end{cases} \tag{3.25}$$

In this case, fundamental solutions are expressed in terms of elementary functions and their integrals. In fact, we find that

$$u = x_3^{-\frac{1}{6}(2r_1+3r_2)} f_0^{\frac{1}{12}(4r_1+3r_2)}(c_1 z + c_2) \tag{3.26}$$

is a solution of the system (3.25), where c_1, c_2 are constants and z is a function such that

$$(\partial_{x_1} z, \partial_{x_2} z, \partial_{x_3} z)$$
$$= \left(\frac{-x_3^{2/3}(2x_1 x_2 + 9x_3)}{f_0}, \frac{(2x_1^2 + 3x_2)x_3^{2/3}}{f_0}, \frac{(-2x_1^2 x_2 - 6x_2^2 + 9x_1 x_3)}{3x_3^{1/3} f_0} \right). \tag{3.27}$$

We are going to explain the outline of the argument how to find the solution (3.26) of the system (3.25).

Putting

$$A_1 = \begin{pmatrix} r_1 & 0 \\ 0 & 2+r_1 \end{pmatrix},$$

$$A_2 = \begin{pmatrix} r_2 x_1 & 0 \\ \frac{1}{3}r_2(2x_1 x_2 + 9x_3) & \frac{1}{3}(-8+3r_2)x_1 \end{pmatrix},$$

$$A_3 = \begin{pmatrix} 0 & 1 \\ -\frac{1}{12}r_2(-4x_2^2 + 3r_2 x_2^2 + 12x_1 x_3) & \frac{1}{3}(-4+3r_2)x_2 \end{pmatrix},$$

we see that (3.25) turns out to be

$$V_i \boldsymbol{u} = A_i \boldsymbol{u} \quad (i = 1, 2, 3), \tag{3.28}$$

where $\boldsymbol{u} = {}^t(u, V_3 u)$. Using the formula (2.11), we define 2×2 matrices B_1, B_2, B_3 from A_1, A_2, A_3. Then (3.28) turns out to be

$$\partial_{x_i} \boldsymbol{u} = B_i \boldsymbol{u} \quad (i = 1, 2, 3). \tag{3.29}$$

To solve (3.29), we introduce $\boldsymbol{v}$ by

$$\boldsymbol{v} = f_0^{-\frac{1}{12}(4r_1+3r_2)} x_3^{\frac{1}{6}(2r_1+3r_2)} \begin{pmatrix} 1 & 0 \\ -\frac{r_2}{2}x_2 x_3^{-2/3} & x_3^{-2/3} \end{pmatrix} \boldsymbol{u}.$$

The system of differential equations for $\boldsymbol{v}$ is then

$$\partial_{x_i} \boldsymbol{v} = \begin{pmatrix} 0 & \varphi_i \\ 0 & 0 \end{pmatrix} \boldsymbol{v} \quad (i = 1, 2, 3), \tag{3.30}$$

where

$$(\varphi_1, \varphi_2, \varphi_3)$$
$$= \left(\frac{-x_3^{2/3}(2x_1x_2 + 9x_3)}{f_0}, \frac{(2x_1^2 + 3x_2)x_3^{2/3}}{f_0}, \frac{(-2x_1^2x_2 - 6x_2^2 + 9x_1x_3)}{3x_3^{1/3}f_0} \right).$$

If $v = {}^t(v_1, v_2)$, then (3.30) is equivalent to the system

$$\begin{cases} \partial_{x_i}v_1 = \varphi_i v_2 & (i = 1, 2, 3), \\ \partial_{x_i}v_2 = 0 & (i = 1, 2, 3). \end{cases} \tag{3.31}$$

At this moment, we note that

$$\partial_{x_i}\varphi_j = \partial_{x_j}\varphi_i \quad (i, j = 1, 2, 3).$$

On the one hand, we find that $u = f_0^{(4r_1+3r_2)/12}x_3^{-(2r_1+3r_2)/6}v_1$. On the other hand, from the second equation of (3.31), it follows that v_2 is a constant. In this manner we have the required result.

<u>Systems of Type II</u>
There are three holonomic systems of Type II.

Type (II.1)

$$\begin{cases} V_1 u = r_1 u, \\ V_2 V_2 u = -\frac{1}{12}r_2\{(-8 + 3r_2)x_1^2 - 12x_2\}u + \frac{1}{3}(-4 + 3r_2)x_1 V_2 u, \\ V_3 u = \frac{1}{4}r_2 x_2^2 u. \end{cases} \tag{3.32}$$

In this case, we find that

$$u = x_3^{-\frac{1}{12}(4r_1+3r_2)}f_0^{\frac{1}{24}(8r_1+3r_2)}(c_1 z + c_2)$$

is a solution of the system (3.32), where c_1, c_2 are constants and z is a function such that

$$(\partial_{x_1}z, \partial_{x_2}z, \partial_{x_3}z)$$
$$= \left(\frac{-x_3^{1/3}(-x_2^2 + 3x_1x_3)}{f_0}, \frac{-x_3^{1/3}(2x_1x_2 + 9x_3)}{2f_0}, \frac{(x_1x_2^2 + 3x_1^2x_3 + 9x_2x_3)}{3x_3^{2/3}f_0} \right). \tag{3.33}$$

The proof is similar to that for the case of Type I.

Type (II.2)

$$\begin{cases} V_1 u = r_1 u, \\ V_2 V_2 u = \frac{1}{12}\{-(64 - 32r_2 + 3r_2^2)x_1^2 + 12r_2 x_2\}u \\ \qquad\quad + \frac{1}{3}(-16 + 3r_2)x_1 V_2 u, \\ V_3 u = \frac{1}{36}\{2(8 - 3r_2)x_1^2 + 3(-16 + 3r_2)x_2\}u + \frac{1}{3}x_1 V_2 u. \end{cases} \tag{3.34}$$

Solutions of (3.34) are expressed in terms of elementary functions. In particular,

$$u_1 = (2x_1^2 + 3x_2)x_3^{\frac{1}{12}(16-4r_1-3r_2)} f_0^{\frac{1}{24}(8r_1+3r_2-24)},$$
$$u_2 = (2x_1x_2 + 9x_3)x_3^{\frac{1}{12}(12-4r_1-3r_2)} f_0^{\frac{1}{24}(8r_1+3r_2-24)}$$

are fundamental solutions of (3.34).

Type (II.3)

$$\begin{cases} V_1 u &= r_1 u, \\ V_2 V_2 u &= -\frac{1}{12}r_2\{(-16 + 3r_2)x_1^2 - 12x_2\}u + \frac{1}{3}(-8 + 3r_2)x_1 V_2 u, \quad (3.35) \\ V_3 u &= \frac{1}{12}r_2(2x_1^2 + 3x_2)u - \frac{1}{3}x_1 V_2 u. \end{cases}$$

In this case, solutions of the system (3.35) are expressed by elementary function. In fact, we find that

$$u_1 = x_2 x_3^{-\frac{1}{12}(4r_1+3r_2)} f_0^{\frac{1}{24}(8r_1+3r_2-8)},$$
$$u_2 = x_3^{-\frac{1}{12}(4r_1+3r_2)} f_0^{\frac{1}{24}(8r_1+3r_2)}$$

are fundamental solutions of the system (3.35).

Remark 3.1. The structure of the fundamental group G of the complement of the Saito free divisor $F_{B,2} = 0$ in $\mathbf{C}^3$ is studied in [7]. In particular, it is shown there that there are three generators a, b, c of G such that a two dimensional representation of G is realized by

$$\rho(a) = u \begin{pmatrix} 1 & \ell^2 \\ 0 & 1 \end{pmatrix}, \quad \rho(b) = v \begin{pmatrix} \ell & 0 \\ 0 & \ell^{-1} \end{pmatrix}, \quad \rho(c) = u \begin{pmatrix} 1 & 1 \\ 0 & 1 \end{pmatrix},$$

where $u, v \in \mathbf{C}^\times$ and $\ell^6 = 1$. The group $\rho(G)$ generated by $\rho(a), \rho(b), \rho(c)$ is solvable. This is a reason why solutions of the holonomic systems in this subsection are expressed by elementary functions.

3.5. The case $F_{B,3}$

The polynomial is

$$F_{B,3} = x_3(2x_2^3 - 9x_1x_2x_3 - 45x_3^2)$$

and its generating matrix is

$$M = \begin{pmatrix} x_1 & 2x_2 & 3x_3 \\ 2x_2 & -\frac{3}{5}(x_1x_2 - 5x_3) & -\frac{6}{5}x_1x_3 \\ 3x_3 & -\frac{3}{5}x_2^2 & -\frac{6}{5}x_2x_3 \end{pmatrix}.$$

There are three holonomic systems. One is of Type I and the remaining two are of Type II. In these cases, solutions of the holonomic systems are expressed by elementary functions.

System of Type I

$$\begin{cases} V_1 u & = r_1 u, \\ V_2 u & = r_2 x_1 u, \\ V_3 V_3 u & = -\frac{1}{5} r_2 (5r_2 - 3) x_2^2 u + \frac{2}{5}(5r_2 - 3) x_2 V_3 u. \end{cases} \tag{3.36}$$

In this case, solutions of (3.36) are expressed by elementary functions. In fact, we find that

$$u_1 = x_3^{-\frac{1}{3}(3r_1 + 10r_2)} (-2x_2^3 + 9x_1 x_2 x_3 + 45x_3^2)^{\frac{1}{3}(2r_1 + 5r_2 - 2)} (-x_2^2 + 3x_1 x_3),$$
$$u_2 = x_3^{-\frac{1}{3}(3r_1 + 10r_2)} (-2x_2^3 + 9x_1 x_2 x_3 + 45x_3^2)^{\frac{1}{3}(2r_1 + 5r_2)}$$

are fundamental solutions of the system (3.36).

Systems of Type II

There are two holonomic systems of Type II.

Type (II.1)

$$\begin{cases} V_1 u & = r_1 u, \\ V_2 V_2 u & = -\frac{1}{20} r_2 \{(5r_2 - 6)x_1^2 - 20x_2\} u + \frac{1}{5}(5r_2 - 3)x_1 V_2 u, \\ V_3 u & = \frac{1}{2} r_2 x_2 u. \end{cases} \tag{3.37}$$

In this case, solutions of (3.37) are expressed by elementary functions. In fact, we find that

$$u_1 = x_2 x_3^{-\frac{1}{3}(3r_1 + 5r_2)} (-2x_2^3 + 9x_1 x_2 x_3 + 45x_3^2)^{\frac{1}{6}(4r_1 + 5r_2 - 2)},$$
$$u_2 = x_3^{-\frac{1}{3}(3r_1 + 5r_2)} (-2x_2^3 + 9x_1 x_2 x_3 + 45x_3^2)^{\frac{1}{6}(4r_1 + 5r_2)}$$

are fundamental solutions of the system (3.37).

Type (II.2)

$$\begin{cases} V_1 u & = r_1 u, \\ V_2 V_2 u & = -\frac{1}{20} r_2 \{(5r_2 - 12)x_1^2 - 20x_2\} u + \frac{1}{5}(5r_2 - 6)x_1 V_2 u, \\ V_3 u & = \frac{1}{20} r_2 (3x_1^2 + 10x_2) u - \frac{3}{10} x_1 V_2 u. \end{cases} \tag{3.38}$$

In this case, solutions of (3.38) are expressed by elementary functions. In fact, we find that

$$u_1 = x_3^{-\frac{1}{3}(3r_1+5r_2)}(-2x_2^3 + 9x_1x_2x_3 + 45x_3^2)^{\frac{1}{6}(4r_1+5r_2-8)}$$
$$\times(-2x_2^4 + 12x_1x_2^2x_3 - 9x_1^2x_3^2 + 60x_2x_3^2)$$
$$u_2 = x_3^{-\frac{1}{3}(3r_1+5r_2)}(-2x_2^3 + 9x_1x_2x_3 + 45x_3^2)^{\frac{1}{6}(4r_1+5r_2)}$$

are fundamental solutions of the system (3.38).

Remark 3.2. The structure of the fundamental group G of the complement of the Saito free divisor $F_{B,3} = 0$ in $\mathbf{C}^3$ is studied in [7]. In particular, it is shown there that $G \simeq \mathbf{Z}^2$. This is a reason why solutions of the holonomic systems in this subsection are expressed by elementary functions.

3.6. *The case* $F_{B,4}$

The polynomial is

$$F_{B,4} = x_3(9x_1^2x_2^2 - 4x_2^3 + 18x_1x_2x_3 + 9x_3^2)$$

and its generating matrix is

$$M = \begin{pmatrix} x_1 & 2x_2 & 3x_3 \\ 2x_2 & 3(3x_1x_2 + x_3) & 6x_1x_3 \\ 3x_3 & 0 & -3x_2x_3 \end{pmatrix}.$$

In this case, there are three holonomic systems of rank two. One is of type I and the remaining two are of Type II. In this subsection, we always put

$$f_0 = 9x_1^2x_2^2 - 4x_2^3 + 18x_1x_2x_3 + 9x_3^2.$$

<u>System of Type I</u>

$$\begin{cases} V_1u = r_1u, \\ V_2u = r_2x_1u, \\ V_3V_3u = -(3r_1 - r_2)(3r_1 - r_2 - 3)x_2^2u - (6r_1 - 2r_2 - 3)x_2V_3u. \end{cases} \quad (3.39)$$

In this case,

$$u = x_3^{\frac{1}{3}(3r_1-r_2)}f_0^{\frac{1}{6}(-2r_1+r_2)}(c_1z + c_2)$$

is a solution of the system (3.39), where c_1, c_2 are constants and z is a function such that

$$(\partial_{x_1}z, \partial_{x_2}z, \partial_{x_3}z) = \left(\frac{(5x_1x_2 + 3x_3)}{x_3f_0^{1/6}}, \frac{2(x_1^2 - x_2)}{x_3f_0^{1/6}}, -\frac{9x_1^2x_2 - 4x_2^2 + 3x_1x_3}{3x_3^2f_0^{1/6}} \right).$$
$$(3.40)$$

The proof is similar to that in subsection 3.4.

Systems of Type II
There are two holonomic systems of Type II.

Type (II.1)

$$\begin{cases} V_1 u & = r_1 u, \\ V_2 V_2 u & = \frac{1}{4}\{(9 - r_3^2)x_1^2 + (-13 + 4r_3 + r_2^2)x_2\}u + r_3 x_1 V_2 u, \qquad (3.41) \\ V_3 u & = -\frac{1}{2}(6r_1 - r_3 + 3)x_2 u. \end{cases}$$

A construction of a solution of (3.41) is accomplished by an argument similar to the case of Type (II.1) in subsection 3.3. To construct solutions of this system, we first put $f_0 = 9(x_3 + x_1 x_2)^2 - 4x_2^3$. Then the singularities of the system are $x_3 = 0$ and $f_0 = 0$. We next define v by

$$v = x_3^{\frac{1}{6}(-6r_1 + r_3 - 3)} f_0^{\frac{1}{12}(4r_1 - r_2 - r_3 + 2)} u.$$

By direct computation, we obtain differential equations for v:

$$\left(\partial_{x_1}^2 + \frac{3(r_2 + 6)x_2(x_1 x_2 + x_3)}{f_0}\partial_{x_1} + \frac{(r_2 + 1)(r_2 + 5)x_2^2}{4f_0}\right)v = 0,$$

$$\left(\partial_{x_3}^2 + \frac{3(r_2 + 6)(x_1 x_2 + x_3)}{f_0}\partial_{x_3} + \frac{(r_2 + 1)(r_2 + 5)}{4f_0}\right)v = 0.$$

Then it is easy to show that

$$v = c(x_2)F\left(\frac{r_2 + 1}{12}, \frac{r_2 + 5}{12}, \frac{1}{2}; \frac{9(x_3 + x_1 x_2)^2}{4x_2^3}\right)$$

is a solution of the two equations for a function $c(x_2)$ of x_2. By the weight condition $V_1 u = r_1 u$, $c(x_2)$ coincides with $x_2^{-\frac{1}{4}(r_2 + 1)}$ up to a constant factor. As a consequence,

$$u = x_3^{\frac{1}{6}(6r_1 - r_3 + 3)} f_0^{\frac{1}{12}(-4r_1 + r_2 + r_3 - 2)} x_2^{-\frac{1}{4}(r_2 + 1)}$$

$$\times F\left(\frac{r_2 + 1}{12}, \frac{r_2 + 5}{12}, \frac{1}{2}; \frac{9(x_3 + x_1 x_2)^2}{4x_2^3}\right)$$

is a solution of the system (3.41).

Type (II.2)

$$\begin{cases} V_1 u & = r_1 u, \\ V_2 V_2 u & = -\frac{1}{4}r_2\{(18 + r_2)x_1^2 - 4x_2\}u + (r_2 + 9)x_1 V_2 u, \qquad (3.42) \\ V_3 u & = \frac{1}{2}\{-3r_2 x_1^2 + (-6r_1 + r_2)x_2\}u + 3x_1 V_2 u. \end{cases}$$

In this case, solutions of (3.42) are expressed by elementary functions and their integrals. In fact,

$$u = x_3^{\frac{1}{6}(6r_1-r_2)} f_0^{\frac{1}{12}(-4r_1+r_2)}(c_1 z + c_2)$$

is a solution of the system (3.42), where c_1, c_2 are constants and z is a function such that

$$(\partial_{x_1} z, \partial_{x_2} z, \partial_{x_3} z)$$
$$= ((15x_1^2 x_2 - 2x_2^2 + 9x_1 x_3) f_0^{1/6} x_3^{-2}, (6x_1^3 - 5x_1 x_2 + 3x_3) f_0^{1/6} x_3^{-2}, \quad (3.43)$$
$$- (9x_1^3 x_2 - 4x_1 x_2^2 + 3x_1^2 x_3 + 2x_2 x_3) f_0^{1/6} x_3^{-3}).$$

The proof is similar to that in subsection 3.4.

3.7. The case $F_{B,6}$

The polynomial is

$$F_{B,6} = 9x_1 x_2^4 + 6x_1^2 x_2^2 x_3 - 4x_2^3 x_3 + x_1^3 x_3^2 - 12x_1 x_2 x_3^2 + 4x_3^3$$

and its generating matrix is

$$M = \begin{pmatrix} x_1 & 2x_2 & 3x_3 \\ 2x_2 & 3x_1 x_2 + \frac{5}{2}x_3 & \frac{9}{2}x_2^2 + \frac{15}{2}x_1 x_3 \\ 3x_3 & \frac{3}{4}(15x_2^2 + x_1 x_3) & 18x_2 x_3 \end{pmatrix}.$$

In this case, there are three holonomic systems of rank two. One is of Type I and the remaining two are of Type II.

System of Type I

$$\begin{cases} V_1 u &= r_1 u, \\ V_2 u &= \frac{1}{12}(20r_1 - 7)x_1 u, \\ V_3 V_3 u &= -\frac{1}{16}(20r_1 - 7)\{2(10r_1 + 1)x_2^2 - 3x_1 x_3\}u + 10(r_1 + 1)x_2 V_3 u. \end{cases}$$
$$(3.44)$$

In this case, the restrictions to $x_1 = 0$ and $x_2 = 0$ of solutions of (3.44) are expressed in terms of hypergeometric functions. We first treat the case $x_1 = 0$. There is a solution u of (3.44) such that

$$u|_{x_1=0} = (x_2^3 - x_3^2)^{\frac{1}{45}(5r_1-4)} x_3^{\frac{1}{45}(5r_1+8)} F\left(\frac{1}{15}, -\frac{2}{15}, \frac{1}{3}; \frac{x_2^3}{x_3^2}\right).$$

We next treat the case $x_2 = 0$. There is a solution u of (3.44) such that

$$u|_{x_2=0} = (x_1^3 + 4x_3)^{\frac{1}{45}(5r_1-4)} x_3^{\frac{1}{45}(10r_1+4)} F\left(-\frac{1}{6}, \frac{13}{30}, \frac{2}{3}; -\frac{x_1^3}{4x_3}\right).$$

Systems of Type II

There are two holonomic systems of Type II.

Type (II.1)

$$\begin{cases} V_1 u = r_1 u, \\ V_2 V_2 u = \frac{1}{144}\{-(20r_1+1)(20r_1+19)x_1^2 + 12(40r_1+11)x_2\}u \\ \qquad\qquad +\frac{5}{3}(2r_1+1)x_1 V_2 u, \\ V_3 u = \frac{1}{4}(20r_1+1)x_2 u. \end{cases} \qquad (3.45)$$

In this case, the restrictions to $x_1 = 0$ and $x_2 = 0$ of solutions of (3.45) are expressed in terms of hypergeometric functions. We first treat the case $x_1 = 0$. There is a solution u of (3.45) such that

$$u|_{x_1=0} = (x_2^3 - x_3^2)^{\frac{1}{45}(5r_1-2)} x_3^{\frac{1}{45}(5r_1+4)} F\left(-\frac{1}{15}, \frac{8}{15}, \frac{2}{3}; \frac{x_2^3}{x_3^2}\right).$$

We next treat the case $x_2 = 0$. There is a solution u of (3.45) such that

$$u|_{x_2=0} = (x_1^3 + 4x_3)^{\frac{1}{45}(5r_1-2)} x_3^{\frac{1}{45}(10r_1+2)} F\left(\frac{1}{6}, -\frac{1}{30}, \frac{1}{3}; -\frac{x_1^3}{4x_3}\right).$$

Type (II.2)

$$\begin{cases} V_1 u = r_1 u, \\ V_2 V_2 u = \frac{1}{144}\{-(20r_1-17)(20r_1+37)x_1^2 + 96(5r_1-2)x_2\}u \\ \qquad\qquad +\frac{5}{3}(2r_1+1)x_1 V_2 u, \\ V_3 u = \frac{1}{16}\{3(-20r_1+17)x_1^2 + 16(5r_1-2)x_2\}u + \frac{9}{4}x_1 V_2 u. \end{cases} \qquad (3.46)$$

In this case, the restrictions to $x_1 = 0$ and $x_2 = 0$ are expressed in terms of hypergeometric functions. We first treat the case $x_1 = 0$. There is a solution u of (3.46) such that

$$u|_{x_1=0} = (x_2^3 - x_3^2)^{\frac{1}{45}(5r_1-11)} x_3^{\frac{1}{45}(5r_1+22)} F\left(-\frac{1}{15}, \frac{2}{15}, \frac{2}{3}; \frac{x_2^3}{x_3^2}\right).$$

We next treat the case $x_2 = 0$. There is a solution u of (3.46) such that

$$u|_{x_2=0} = (x_1^3 + 4x_3)^{\frac{1}{45}(5r_1-11)} x_3^{\frac{1}{45}(10r_1+11)} F\left(\frac{1}{6}, -\frac{13}{30}, \frac{1}{3}; -\frac{x_1^3}{4x_3}\right).$$

3.8. *The case $F_{H,1}$*

The polynomial is

$$F_{H,1} = -8x_1^9 x_2^2 - 20x_1^6 x_2^3 - 230x_1^3 x_2^4 - 135x_2^5 + 8x_1^7 x_2 x_3 + 120x_1^4 x_2^2 x_3 \\ +450x_1 x_2^3 x_3 + 8x_1^5 x_3^2 - 100x_1^2 x_2 x_3^2 - 100x_3^3$$

and its generating matrix is

$$M = \begin{pmatrix} x_1 & 3x_2 & 5x_3 \\ 3x_2 & 2x_3 + 2x_1^2 x_2 & 7x_1 x_2^2 + 2x_1^4 x_2 \\ 5x_3 & 7x_1 x_2^2 + 2x_1^4 x_2 & \frac{1}{2}(15x_2^3 + 4x_1^4 x_3 + 18x_1^3 x_2^2) \end{pmatrix}.$$

Note that $F_{H,1}$ is regarded as the discriminant of the reflection group of type H_3.

In this case, there are three holonomic systems of rank two. One is of Type I and the remaining two are of Type II.

System of Type I

$$\begin{cases} V_1 u &= r_1 u, \\ V_2 u &= \frac{2}{15}(r_1 + 2)x_1^2 u, \\ V_3 V_3 u = &-\frac{2}{225}\begin{pmatrix} 8(2+r_1)^2 x_1^8 + 10(2+r_1)(-7+4r_1)x_1^5 x_2 \\ +25(-4+r_1)(-5+2r_1)x_1^2 x_2^2 \\ -300(1+2r_1)x_1^3 x_3 - 375(2+r_1)x_2 x_3 \end{pmatrix} u \\ &+\frac{4}{15}(2+r_1)x_1(2x_1^3 + 5x_2)V_3 u. \end{cases} \tag{3.47}$$

In this case, the restrictions to $x_1 = 0$ and $x_2 = 0$ of solutions of (3.47) are expressed in terms of hypergeometric function. We first treat the case $x_1 = 0$. There is a solution u of (3.47) such that

$$u|_{x_1=0} = (27x_2^5 + 20x_3^3)^{\frac{1}{15}(r_1+2)} x_3^{-\frac{2}{5}} F\left(\frac{2}{15}, \frac{7}{15}, \frac{3}{5}; -\frac{27x_2^5}{20x_3^3}\right).$$

We next treat the case $x_2 = 0$. There is a solution u of (3.47) such that

$$u|_{x_2=0} = (2x_1^5 - 25x_3)^{\frac{1}{15}(r_1+2)} x_3^{\frac{1}{15}(2r_1-2)} F\left(\frac{2}{5}, \frac{2}{5}, \frac{4}{5}; \frac{2x_1^5}{25x_3}\right).$$

Systems of Type II
There are two systems of Type II.

Type (II.1)

$$\begin{cases} V_1 u &= r_1 u, \\ V_2 V_2 u = &-\frac{4}{225}x_1\{(r_1^2 + 2r_1 - 44)x_1^3 - 45(r_1 + 12)x_2\}u \\ &+\frac{4}{15}(r_1 + 1)x_1^2 V_2 u, \\ V_3 u &= \frac{2}{15}x_1\{2(r_1 + 1)x_1^3 + 5(r_1 + 4)x_2\}u. \end{cases} \tag{3.48}$$

In this case, the restriction to $x_1 = 0$ of any solution of (3.48) is expressed in terms of hypergeometric functions. In particular, there is a solution u of (3.48) such that

$$u|_{x_1=0} = (27x_2^5 + 20x_3^3)^{\frac{1}{15}(r_1-2)}x_3^{\frac{2}{5}}F\left(-\frac{2}{15}, \frac{8}{15}, \frac{4}{5}; -\frac{27x_2^5}{20x_3^3}\right).$$

On the other hand, we obtain differential equations

$$\partial_{x_1}\begin{pmatrix} u \\ V_2u \end{pmatrix} = \frac{1}{3(2x_1^5-25x_3)}\begin{pmatrix} 2(r_1-2)x_1^4 & 0 \\ -20(r_1+4)x_1x_3 & 2(r_1+4)x_1^4 \end{pmatrix}\begin{pmatrix} u \\ V_2u \end{pmatrix}$$

$$\partial_{x_3}\begin{pmatrix} u \\ V_2u \end{pmatrix} = \frac{1}{15x_3(2x_1^5-25x_3)}\begin{pmatrix} \begin{pmatrix} 4(r_1+1)x_1^5 \\ -75r_1x_3 \end{pmatrix} & 0 \\ 20(r_1+4)x_1^2x_3 & \begin{pmatrix} (r_1+1)x_1^5 \\ -75(r_1+2)x_3 \end{pmatrix} \end{pmatrix}\begin{pmatrix} u \\ V_2u \end{pmatrix}$$

by restricting the system (3.48) to $x_2 = 0$. Putting

$$h = x_3^{\frac{2}{15}(r_1+1)}(25x_3 - 2x_1^5)^{\frac{1}{15}(r_1-2)},$$

we find that if u is a solution of (3.48), then

$$u|_{x_2=0} = c_1h, \quad (V_2u)|_{x_2=0} = \left(\frac{2}{15}c_1(r_1+4)x_1^2 + c_2(2x_1^5 - 25x_3)^{\frac{2}{5}}\right)h,$$

where c_1, c_2 are constants.

Type (II.2)

$$\begin{cases} V_1u = r_1u, \\ V_2V_2u = -\frac{2}{225}x_1\{2(r_1+1)^2x_1^3 - 45(2r_1-3)x_2\}u \\ \qquad\qquad + \frac{4}{15}(r_1+1)x_1^2V_2u, \\ V_3u = \frac{1}{15}x_1\{2(r_1+1)x_1^3 + 5(2r_1-1)x_2\}u + x_1^2V_2u. \end{cases} \tag{3.49}$$

In this case, the restrictions to $x_1 = 0$ and $x_2 = 0$ of solutions of (3.49) are expressed in terms of hypergeometric functions. We first treat the case $x_1 = 0$. There is a solution u of (3.49) such that

$$u|_{x_1=0} = (27x_2^5 + 20x_3^3)^{\frac{1}{15}(r_1+1)}x_3^{-\frac{1}{5}}F\left(\frac{1}{15}, \frac{11}{15}, \frac{4}{5}; -\frac{27x_2^5}{20x_3^3}\right).$$

We next treat the case $x_2 = 0$. There is a solution u of (3.49) such that

$$u|_{x_2=0} = (2x_1^5 - 25x_3)^{\frac{1}{15}(r_1+1)}x_3^{\frac{1}{15}(2r_1-1)}F\left(\frac{1}{5}, \frac{1}{5}, \frac{2}{5}; \frac{2x_1^5}{25x_3}\right).$$

3.9. *The case $F_{H,2}$*

The polynomial is

$$F_{H,2} = 100x_1^3x_2^4 + x_2^5 + 40x_1^4x_2^2x_3 - 10x_1x_2^3x_3 + 4x_1^5x_3^2 - 15x_1^2x_2x_3^2 + x_3^3$$

and its generating matrix is

$$M = \begin{pmatrix} x_1 & 3x_2 & 5x_3 \\ 3x_2 & 36x_1^2x_2 + 6x_3 & 90x_1x_2^2 + 90x_1^2x_3 \\ 5x_3 - \frac{10}{3}(12x_1^3 - 55x_2)x_1x_2 & -\frac{50}{3}(6x_1^3x_2^2 - x_2^3 + 6x_1^4x_3 - 18x_1x_2x_3) \end{pmatrix}.$$

In this case, there are two holonomic systems of rank two. One is of Type I and the other is of Type II.

System of Type I

$$\begin{cases} V_1 u & = r_1 u, \\ V_2 u & = 3(4r_1 - 1)x_1^2 u, \\ V_3 V_3 u = -\frac{25}{9} \begin{pmatrix} 4(-1 + 4r_1)(17 + 4r_1)x_1^8 \\ -12(-27 + 100r_1 + 40r_1^2)x_1^5x_2 \\ +15(-7 + 20r_1 + 60r_1^2)x_1^2x_2^2 \\ +8(12r_1 - 1)x_1^3x_3 - 9(10r_1 - 1)x_2x_3 \end{pmatrix} u \\ \qquad\qquad -\frac{20}{3}(2 + r_1)x_1(4x_1^3 - 15x_2)V_3 u. \end{cases} \tag{3.50}$$

In this case, the restrictions to $x_1 = 0$ and $x_2 = 0$ of solutions of (3.50) are expressed in terms of hypergeometric functions. We first treat the case $x_1 = 0$. There is a solution u of (3.50) such that

$$u|_{x_1=0} = (x_2^5 + x_3^3)^{\frac{1}{30}(2r_1-1)} x_3^{\frac{1}{10}} F\left(-\frac{1}{30}, \frac{3}{10}, \frac{3}{5}; -\frac{x_2^5}{x_3^3}\right).$$

We next treat the case $x_2 = 0$. There is a solution u of (3.50) such that

$$u|_{x_2=0} = (4x_1^5 + x_3)^{\frac{1}{30}(2r_1-1)} x_3^{\frac{1}{30}(4r_1+1)} F\left(-\frac{1}{15}, \frac{8}{15}, \frac{4}{5}; -\frac{4x_1^5}{x_3}\right).$$

System of Type II

$$\begin{cases} V_1 u & = r_1 u, \\ V_2 V_2 u = -9(4r_1 + 7)x_1\{(4r_1 + 1)x_1^3 - 2x_2\}u + 24(r_1 + 1)x_1^2 V_2 u, \\ V_3 u & = \frac{5}{3}(30r_1 + 7)x_1x_2 u - \frac{10}{9}x_1^2 V_2 u. \end{cases} \tag{3.51}$$

In this case, the restrictions to $x_1 = 0$ and $x_2 = 0$ of solutions of (3.51) are expressed in terms of hypergeometric functions. We first treat the case

$x_1 = 0$. There is a solution u of (3.51) such that

$$u|_{x_1=0} = (x_2^5 + x_3^3)^{\frac{1}{30}(2r_1-3)} x_3^{\frac{3}{10}} F\left(-\frac{1}{10}, \frac{17}{30}, \frac{4}{5}; -\frac{x_2^5}{x_3^3}\right).$$

We next treat the case $x_2 = 0$. There is a solution u of (3.51) such that

$$u|_{x_2=0} = (4x_1^5 + x_3^3)^{\frac{1}{30}(2r_1-3)} x_3^{\frac{1}{30}(4r_1+3)} F\left(-\frac{1}{15}, \frac{2}{15}, \frac{2}{5}; -\frac{4x_1^5}{x_3}\right).$$

3.10. The case $F_{H,3}$

The polynomial is

$$F_{H,3} = 8x_1^3 x_2^4 + 108x_2^5 - 36x_1 x_2^3 x_3 - x_1^2 x_2 x_3^2 + 4x_3^3$$

and its generating matrix is

$$M = \begin{pmatrix} x_1 & 3x_2 & 5x_3 \\ 3x_2 & \frac{1}{10}(x_1^2 x_2 + 2x_3) & \frac{23}{10}x_1 x_2^2 + \frac{3}{20}x_1^2 x_3 \\ 5x_3 & 5x_1 x_2^2 & \frac{15}{2}x_2(2x_2^2 + x_1 x_3) \end{pmatrix}.$$

In this case, there are two holonomic systems of rank two. One is of Type I and the other is of Type II.

System of Type I

$$\begin{cases} V_1 u & = r_1 u, \\ V_2 u & = \frac{1}{600}(16r_1 - 13)x_1^2 u, \\ V_3 V_3 u & = -\frac{16r_1-13}{144}x_2\{(16r_1 + 17)x_1^2 x_2 - 60x_3\}u + \frac{8}{3}(r_1 + 2)x_1 x_2 V_3 u. \end{cases}$$

(3.52)

In this case, the restriction to $x_1 = 0$ of any solution of (3.52) is expressed in terms of hypergeometric functions. In particular, there is a solution u of (3.52) such that

$$u|_{x_1=0} = (27x_2^5 + x_3^3)^{\frac{1}{30}(2r_1-5)} x_3^{\frac{1}{2}} F\left(\frac{1}{6}, -\frac{1}{6}, \frac{3}{5}; -\frac{27x_2^5}{x_3^3}\right).$$

On the other hand, we obtain differential equations by restricting (3.52) to $x_2 = 0$:

$$\partial_{x_1}\begin{pmatrix} u \\ V_3 u \end{pmatrix} = \begin{pmatrix} 0 & \frac{1}{5x_3} \\ 0 & 0 \end{pmatrix}\begin{pmatrix} u \\ V_3 u \end{pmatrix},$$

$$\partial_{x_3}\begin{pmatrix} u \\ V_3 u \end{pmatrix} = \begin{pmatrix} \frac{r_1}{5x_3} & -\frac{x_1}{25x_3^2} \\ 0 & \frac{r_1+4}{5x_3} \end{pmatrix}\begin{pmatrix} u \\ V_3 u \end{pmatrix}.$$

Solving these equations, we find that if u is a solution of (3.52), then

$$u|_{x_2=0} = c_1 x_3^{\frac{r_1}{5}} + c_2 x_1 x_3^{\frac{1}{5}(r_1-1)}, \quad (V_3 u)|_{x_2=0} = 5 c_2 x_3^{\frac{1}{5}(r_1+4)},$$

where c_1, c_2 are constants.

System of Type II

$$
\begin{cases}
V_1 u &= r_1 u, \\
V_2 V_2 u &= -\frac{1}{360000} x_1 \{(16 r_1 + 1)(16 r_1 + 31) x_1^3 - 3600(16 r_1 + 3) x_2\} u \\
&\quad + \frac{4}{75}(r_1 + 1) x_1^2 V_2 u, \\
V_3 u &= \frac{1}{12}(16 r_1 + 1) x_1 x_2 u.
\end{cases}
$$

(3.53)

In this case, the restriction to $x_1 = 0$ is expressed in terms of hypergeometric functions. There is a solution u of (3.53) such that

$$u|_{x_1=0} = (27 x_2^5 + x_3^3)^{\frac{1}{30}(2 r_1 - 1)} x_3^{\frac{1}{10}} F\left(-\frac{1}{30}, \frac{19}{30}, \frac{4}{5}; -\frac{27 x_2^5}{x_3^3}\right).$$

On the other hand, we obtain differential equations by restricting (3.53) to $x_2 = 0$:

$$
\partial_{x_1}\begin{pmatrix} u \\ V_2 u \end{pmatrix} = \begin{pmatrix} 0 & 0 \\ \frac{16 r_1 + 1}{300} x_1 & 0 \end{pmatrix}\begin{pmatrix} u \\ V_2 u \end{pmatrix},
$$

$$
\partial_{x_3}\begin{pmatrix} u \\ V_2 u \end{pmatrix} = \frac{1}{5 x_3}\begin{pmatrix} r_1 & 0 \\ -\frac{16 r_1 + 1}{300} x_1^2 & r_1 + 2 \end{pmatrix}\begin{pmatrix} u \\ V_2 u \end{pmatrix}.
$$

Solving these equations, we find that

$$u|_{x_2=0} = c_1 x_3^{\frac{r_1}{5}}, \quad (V_2 u)|_{x_2=0} = \frac{16 r_1 + 1}{600} c_1 x_1^2 x_3^{\frac{r_1}{5}} + c_2 x_3^{\frac{1}{5}(r_1+2)}$$

is the restriction of a solution of (3.53) to $x_2 = 0$, where c_1, c_2 are constants.

4. Relationship with algebraic solutions of Painlevé sixth equation

The polynomial $F_{A,2}$ was found by M. Sato about forty years ago. This polynomial looks like the discriminant of the reflection group of type A_3. The motivation of the paper [8] is to find weighted homogeneous polynomials of three variables like $F_{A,2}$. As a result, the author found seventeen polynomials including discriminants of irreducible real reflection groups of rank three. The most important property of them is that they define Saito free divisors. As the next stage of the study following the paper [4], the author investigated systems of uniformization equations with singularities along the Saito free divisor of simple type. The result was summarized in [9]. The existence of such a system is deeply related with that of three dimensional representation of the fundamental group of the complement of the Saito free divisor. The construction of such a system is reduced to an integrability condition which one needs a simple but complicated computation to solve. At this moment, it is underlined that in [4], Saito used a suggestive argument to construct solutions of a special case of systems of uniformization equations. In the special case he treated, the system has a quotient in the sense of $\mathcal{D}$-module and the quotient is a holonomic system of rank two and is of Type I in our terminology. This leads the author to the study on the construction of holonomic systems of Type I and Type II.

It is a basic problem whether holonomic systems obtained in the previous section are reduced to elementary systems or classically known systems or not. For example, take holonomic systems given in subsection 3.4. Their solutions are expressed by elementary functions, namely by product of complex powers of polynomials and their integrals. On the other hand, take the holonomic system given in subsection 3.2. In this case, their solutions are expressed by Gaussian hypergeometric functions. This means that the holonomic system in subsection 3.2 is reduced to an ordinary differential equation of hypergeometric type. But in some of holonomic systems obtained in the previous section, it seems hard to show whether solutions of them are expressed by elementary functions or special functions or not. We treat the holonomic system of Type I in subsection 3.9 and show that it induces an ordinary differential equation of second order with three singular points and an apparent singular point. This is an evidence that its solutions are not expressed by neither elementary functions nor hypergeometric functions.

Put $F = F_{H,2}$ and consider the system (3.50) (cf. subsection 3.9). Then by direct computation, we obtain a differential equation for u with respect to x_3:

$$\partial_{x_3}^2 u + \frac{P_1(x)}{(x_3 - a_s)F} \partial_{x_3} u + \frac{P_2(x)}{(x_3 - a_s)F^2} u = 0, \qquad (4.54)$$

where P_1, P_2 are polynomials of x_3 with rational coefficients of x_1, x_2 and a_s is a rational function of x_1, x_2. We now put $u = F^{(2r_1-1)/30} y$ and obtain a differential equation for y from (4.54). Then

$$\partial_{x_3}^2 y + \left(\frac{2}{3} \frac{\partial_{x_3} F}{F} - \frac{1}{x_3 - a_s} \right) \partial_{x_3} y + \left(\frac{c_0 x_3^2 + c_1 x_3 + c_2}{F} - \frac{c_0}{x_3 - a_s} \right) y = 0,$$

where

$$c_0 = \frac{x_1}{15x_2(12x_1^3 - x_2)},$$
$$c_1 = \frac{80x_1^6 - 456x_1^3 x_2 + 33x_2^2}{300x_2(12x_1^3 - x_2)},$$
$$c_2 = -\frac{x_1^2(528x_1^6 - 3224x_1^3 x_2 + 825x_2^2)}{900(12x_1^3 - x_2)},$$
$$a_s = -\frac{3(4x_1^3 x_2 - x_2^2)}{2x_1}.$$

Let z_0, z_1, z_2 be solutions of $F = 0$ as a cubic polynomial of x_3, namely,

$$F = (x_3 - z_0)(x_3 - z_1)(x_3 - z_2).$$

Using a_s, z_0, z_1, z_2, we put

$$t = \frac{z_2 - z_0}{z_1 - z_0}, \quad w = \frac{a_s - z_0}{z_1 - z_0}.$$

Then both t, w are algebraic functions of $m = \frac{x_2}{x_1^3}$. Putting

$$m = \frac{s^2(s+1)^2}{(s^2 + s + 1)^2},$$

we find that

$$t = \frac{s^5(s+2)(2s^2 + 3s + 3)^2}{(2s+1)(3s^2 + 3s + 2)^2}, \quad w = \frac{s^2(s+2)(s^2 + 1)(2s^2 + 3s + 3)}{2(s^2 + s + 1)(3s^2 + 3s + 2)}.$$

The pair (t, w) is an algebraic solution of Painlevé sixth equation given in [2], p.23. This suggests that solutions of the differential equation (3.50) are not expressed by special functions.

For the details of the argument above, see [10].

Remark 4.1. The author thanks M. Kato of Univ. of Ryukyus for explaining him an idea of the argument on deriving an algebraic solution of Painlevé sixth equation from the systems of equations obtained in this paper.

References

1. A. G. Aleksandrov, Moduli of logarithmic connections along Saito free divisor, Contemp. Math., **314** (2002), 2-23.
2. P. Boalch, The fifty-two icosahedral solutions to Painlevé VI, J. Reine Angew. Math., **596**, (2006), 183-214.
3. M. Kato and J. Sekiguchi, Systems of uniformization equations with respect to the discriminant sets of complex reflection groups of rank three. Preprint.
4. K. Saito, On the uniformization of complements of discriminant loci. RIMS Kôkyuroku **287** (1977), 117-137.

5. K. Saito, Theory of logarithmic differential forms and logarithmic vector fields. J. Faculty of Sciences, Univ. Tokyo **27** (1980), 265-291.
6. K. Saito, Uniformization of orbifold of a finite reflection group. In *"Frobenius Manifold, Quantum Cohomology and Singularities."* A Publication of the Max-Planck-Institute, Mathematics, Bonn, 265-320.
7. K. Saito and T. Ishibe, Monoids in the fundamental groups of the complement of logarithmic Saito free divisors in $\mathbf{C}^3$, J. Algebra **344** (2011), 137-160.
8. J. Sekiguchi, A classification of weighted homogeneous Saito free divisors, J. Math. Soc. Japan, **61** (2009), 1071-1095.
9. J. Sekiguchi, Systems of uniformization equations along Saito free divisors and related topics. In *"The Third Japanese-Australian Workshop on Real and Complex Singularities"*, Proceedings of the Centre for Mathematics and its Applications, **43** (2010), 83-126.
10. J. Sekiguchi, Free divisors, holonomic systems and algebraic solutions of Painlevé sixth equation. In preparation.

Parametric local cohomology classes and Tjurina stratifications for μ-constant deformations of quasi-homogeneous singularities

Shinichi Tajima

Division of Mathematics, University of Tsukuba, Tsukuba, Ibaraki 305-8571, Japan
tajima@math.tsukuba.ac.jp

Local cohomology classes attached to semi-quasihomogeneous hypersurface isolated singularities are considered. A new effective method to compute Tjurina stratifications associated with $\mu-$constant deformation $f_t(x) = F(x,t)$, $t \in T$ of weighted homogeneous isolated singularities is proposed. The proposed method also computes on each stratum, via Grothendieck local duality, a parametric standard basis of the relevant ideal quotient $J_t : f_t$, where J_t stands for the Jacobi ideal of the function f_t in the local ring of germs of holomorphic functions. The key idea in this approach is the use of parametric local cohomology classes.

Keywords: standard basis, Grothendieck local duality, algebraic local cohomology

AMS classification numbers: 32S25, 32S30, 32C36

1. Introduction

Let X be an open neighborhood of the origin O of the n-dimensional complex space $\mathbb{C}^n$ with coordinates $x = (x_1, x_2, ..., x_n)$ and let $\mathcal{O}_X$ be the sheaf on X of holomorphic functions. Let f be a holomorphic function defined on X with an isolated singularity at the origin O and let J_f denote the Jacobi ideal $(\frac{\partial f}{\partial x_1}, \frac{\partial f}{\partial x_2}, ..., \frac{\partial f}{\partial x_n})$ in $\mathcal{O}_{X,O}$ generated by the partial derivatives $\frac{\partial f}{\partial x_1}, \frac{\partial f}{\partial x_2}, ..., \frac{\partial f}{\partial x_n}$, where $\mathcal{O}_{X,O}$ is the stalk at O of the sheaf $\mathcal{O}_X$. Let (f, J_f) denote the ideal $(f, \frac{\partial f}{\partial x_1}, \frac{\partial f}{\partial x_2}, ..., \frac{\partial f}{\partial x_n})$ in $\mathcal{O}_{X,O}$ generated by f and $\frac{\partial f}{\partial x_1}, \frac{\partial f}{\partial x_2}, ..., \frac{\partial f}{\partial x_n}$.

The Milnor number $\mu_f = \dim_{\mathbb{C}}(\mathcal{O}_{X,O}/J_f)$ is a topological invariant and the Tjurina number τ_f introduced by Tjurina in the context of deformation theory, which is defined to be the dimension of the vector space $\mathcal{O}_{X,O}/(f, J_f)$, is an analytic invariant of the singularity. Tjurina numbers have been extensively studied ([17], [18], [28], [39]). It turned out that the Tjurina number is closely related with several complex analytic properties of singularities ([4], [12], [29], [37], [39]). In 1989, B. Martin and G. Pfister ([22]) constructed an algorithm of computing parameter dependency of Tjurina numbers of μ-constant deformations of quasi-homogeneous hypersurface isolated singularities. The algorithm has been successfully applied to study of semi quasi-homogeneous singularities ([6], [7]).

In this paper, we propose an alternative approach, in a context of computational algebraic analysis, to compute Tjurina stratifications of μ-constant deformations. The resulting algorithm has already been implemented in a computer algebra system Risa/Asir ([27]). A description of the algorithm will appear elsewhere. Note that the method is extendable to handle μ-constant deformations of Newton non-degenerate hypersurface isolated singularities.

Note also that the framework we adopted is expected to provide an effective method to study certain families of holonomic D_X-modules attached to μ-constant deformations of hypersurface singularities ([37]).

2. Local cohomology and non-quasihomogeneity

Let $\mathcal{H}^n_{\{O\}}(\Omega^n_X)$ be the local cohomology supported at the origin O, where Ω^n_X is the sheaf on X of holomorphic differential n-forms. Let $\mathcal{H}^n_{[O]}(\Omega^n_X)$ denote the algebraic local cohomology supported at the origin O defined by

$$\mathcal{H}^n_{[O]}(\Omega^n_X) = \lim_{k \to \infty} \mathcal{E}xt^n_{\mathcal{O}_X}(\mathcal{O}_X/(x_1, x_2, ..., x_n)^k, \Omega^n_X),$$

where $(x_1, x_2, ..., x_n)$ is the maximal ideal generated by $x_1, x_2, ..., x_n$. Then, the space $\mathcal{H}^n_{\{O\}}(\Omega^n_X)$ naturally has a structure of a Fréchet-Schwartz topological vector space and the space $\mathcal{H}^n_{[O]}(\Omega^n_X)$ has a structure of the dual Fréchet-Schwartz topological vector space ([14], [31], [38]).

A theory of Functional Analysis asserts that the topological vector space $\mathcal{H}^n_{\{O\}}(\Omega^n_X)$ and the space $\mathcal{O}_{X,O}$ of convergent power series at the origin are mutually strong dual via the local residue pairing ([14], [15], [23]). The same duality holds between the space $\mathcal{H}^n_{[O]}(\Omega^n_X)$ and the space $\hat{\mathcal{O}}_{X,O}$ of formal power series at the origin ([16], [19]).

Set

$$W_{J_f} = \{\omega \in \mathcal{H}^n_{\{O\}}(\Omega^n_X) \mid \frac{\partial f}{\partial x_1}\omega = \frac{\partial f}{\partial x_2}\omega = \cdots = \frac{\partial f}{\partial x_n}\omega = 0\}.$$

Since f has isolated singularity at the origin, W_f coincides with the space $\widehat{W}_f$ of the algebraic local cohomology classes defined by

$$\widehat{W}_{J_f} = \{\omega \in \mathcal{H}^n_{[O]}(\Omega^n_X) \mid \frac{\partial f}{\partial x_1}\omega = \frac{\partial f}{\partial x_2}\omega = \cdots = \frac{\partial f}{\partial x_n}\omega = 0\}.$$

Note that an efficient algorithm for computing a basis of the vector space $\widehat{W}_{J_f}$ is described in [34], [35].

Let J_f (resp. $\hat{J}_f$) denote the Jacobi ideal generated by $\frac{\partial f}{\partial x_1}, \frac{\partial f}{\partial x_2} \cdots \frac{\partial f}{\partial x_n}$ in the local ring $\mathcal{O}_{X,O}$ (resp. $\hat{\mathcal{O}}_{X,O}$). Then, the local duality theorem ([13], [15]) asserts that there is a non-degenerate pairing

$$\mathrm{res}_O(\ ,\ ) : \mathcal{O}_{X,O}/J_f \times W_{J_f} \longrightarrow \mathbb{C}.$$

An efficient method to evaluate the pairing above is also given in [34], [35]. The non-degeneracy of the pairing implies in particular the fact that, a convergent power series $h \in \mathcal{O}_{X,O}$ is in the ideal J_f if and only if $\mathrm{res}_O(h, \omega) = 0$ for every $\omega \in W_{J_f}$. Thus, the Jacobi ideal J_f is completely determined by the space $W_{J_f}(= \widehat{W}_{J_f})$ of (algebraic) local cohomology classes via local residues ([26], [33], [34]).

Note that, since $W_{J_f} = \mathcal{E}xt^n_{\mathcal{O}_X}(\mathcal{O}_X/J_f, \Omega^n_X)$, the non-degenerate pairing above coincides with the classical pairing

$$\mathcal{O}_{X,O}/J_f \times \mathcal{E}xt^n_{\mathcal{O}_X}(\mathcal{O}_X/J_f, (\Omega^n_X)) \longrightarrow \mathbb{C}$$

induced from Yoneda pairing in homological algebra. The corresponding non-degeneracy coincides with an analytical version of the Grothendieck local duality theorem on residues ([8], [9], [10], [11], [16], [19]).

Now we introduce another vector space W_{T_f} defined to be the set of local cohomology classes in $\mathcal{H}^n_{\{O\}}(\Omega^n_X)$ that are annihilated by the ideal generated by f and $\frac{\partial f}{\partial x_1}, \frac{\partial f}{\partial x_2} \cdots \frac{\partial f}{\partial x_n}$ in the local ring $\mathcal{O}_{X,O}$:

$$W_{T_f} = \{\omega \in \mathcal{H}^n_{\{O\}}(\Omega^n_X) \mid f\omega = \frac{\partial f}{\partial x_1}\omega = \frac{\partial f}{\partial x_2}\omega = \cdots = \frac{\partial f}{\partial x_n}\omega = 0\},$$

that is also equal to

$$\widehat{W}_{T_f} = \{\omega \in \mathcal{H}^n_{[O]}(\Omega^n_X) \mid f\omega = \frac{\partial f}{\partial x_1}\omega = \frac{\partial f}{\partial x_2}\omega = \cdots = \frac{\partial f}{\partial x_n}\omega = 0\}.$$

It follows directly from the definitions that $W_{T_f} = \{\omega \in W_{J_f} \mid f\omega = 0\}$. Note also that the dimension of the vector space W_{J_f} is equal to the Milnor number μ_f defined to be $\mu_f = \dim_{\mathbb{C}}(\mathcal{O}_{X,O}/J_f)$ and the dimension of the vector space W_{T_f} is equal to the Tjurina number $\tau_f = \dim_{\mathbb{C}}(\mathcal{O}_{X,O}/(f, J_f))$

Let $\varphi : W_{J_f} \longrightarrow W_{J_f}$ be a map defined by $\varphi(\omega) = f\omega$ and set $W_{Q_f} = \{f\omega \mid \omega \in W_{J_f}\}$. Since $\mathrm{Ker}\varphi = W_{T_f}$ and $\mathrm{Im}\varphi = W_{Q_f}$, we have the following result.

Lemma 1 *The sequence*

$$0 \longrightarrow W_{T_f} \longrightarrow W_{J_f} \longrightarrow W_{Q_f} \longrightarrow 0$$

of vector spaces defined by φ is exact.

Theorem 2

(i) $\dim W_{Q_f} = \mu_f - \tau_f$.

(ii) $Ann_{\mathcal{O}_{X,O}}(W_{Q_f}) = \{h \in \mathcal{O}_{X,O} \mid hf \in J_f\}$.

Proofs (i) Trivial. (ii) See the next section.

Now let $U_0 = X$ and $U_j = \{x = (x_1, \ldots, x_n) \in X \mid x_j \neq 0\}$ for $j = 1, \ldots, n$. Consider the pair $(X, X - \{O\})$ of open sets and its relative covering $(\mathcal{U}, \mathcal{U}')$ where $\mathcal{U} = \{U_0, U_1, \ldots, U_n\}$ and $\mathcal{U}' = \{U_1, \ldots, U_n\}$. Then, any local cohomology class in the space W_{J_f}, which is actually a subspace of $\mathcal{H}^n_{[O]}(\Omega^n_X)$, can be represented as a finite sum of the form

$$\sum_\lambda c_\lambda \left[\frac{1}{x^\lambda} dx \right] = \sum c_{(l_1 \ldots l_n)} \left[\frac{1}{x_1^{l_1} \ldots x_n^{l_n}} dx \right],$$

where $c_\lambda \in \mathbb{C}$ with $\lambda = (l_1, \ldots, l_n) \in \mathbb{N}^n_+$, $dx = dx_1 \wedge dx_2 \wedge \cdots \wedge dx_n$ and $[\frac{1}{x^\lambda} dx]$ is a relative Čech cohomology class ([26], [30], [34], [35]).

We also use the notation $(\sum c_\lambda [\frac{1}{x^\lambda}]) dx$ or $[\sum c_\lambda \frac{1}{x^\lambda}] dx$ for representing algebraic local cohomology classes in $\mathcal{H}^n_{[O]}(\Omega^n_X)$.

Note that

$$x^\kappa \left[\frac{1}{x^\lambda} dx \right] = \begin{cases} [\frac{1}{x^{\lambda - \kappa}} dx], & l_i > k_i, i = 1, \ldots, n \\ 0, & \text{otherwise}, \end{cases}$$

where $\kappa = (k_1, \ldots, k_n) \in \mathbb{N}^n$, $\lambda = (l_1, \ldots, l_n) \in \mathbb{N}^n_+$, and $\lambda - \kappa = (l_1 - k_1, \ldots, l_n - k_n)$.

Example 1 (E_{12} singularity) Let $f(x, y) = x^3 + y^7 + axy^5$.

The Milnor number μ_f is 12 and the following 12 algebraic local cohomology classes constitute a basis of the vector space W_{J_f} ([33]).

$$[\frac{dx \wedge dy}{xy}], [\frac{dx \wedge dy}{xy^2}], [\frac{dx \wedge dy}{x^2 y}], [\frac{dx \wedge dy}{xy^3}], [\frac{dx \wedge dy}{x^2 y^2}]$$

$$[\frac{dx \wedge dy}{xy^4}], [\frac{dx \wedge dy}{x^2 y^3}], [\frac{dx \wedge dy}{xy^5}], [\frac{dx \wedge dy}{x^2 y^4}],$$

$$([\frac{1}{xy^6}] - \frac{a}{3}[\frac{1}{x^3 y}]) dx \wedge dy,$$

$$([\frac{1}{x^2y^5}] - \frac{5a}{7}[\frac{1}{xy^7}] + \frac{5a}{21}[\frac{1}{x^3y^2}])dx \wedge dy,$$

and

$$([\frac{1}{x^2y^6}] - \frac{5a}{7}[\frac{1}{xy^8}] - \frac{a}{3}[\frac{1}{x^4y}] + \frac{5a^2}{21}[\frac{1}{x^3y^3}])dx \wedge dy.$$

It is easy to see that the first 11 local cohomology classes in the list above belong to W_{T_f}. We have by a direct computation

$$W_{Q_f} = \text{Span}_{\mathbb{C}}\{a[\frac{dx \wedge dy}{xy}]\},$$

which implies in particular that $\tau_f = 12 - 1 = 11$, if $a \neq 0$.

Example 2 Let $f(x, y) = x^3 + y^7 + bxy^6$.

The following 12 algebraic local cohomology classes constitute a basis of the vector space W_{J_f}.

$$[\frac{dx \wedge dy}{xy}], [\frac{dx \wedge dy}{xy^2}], [\frac{dx \wedge dy}{x^2y}], [\frac{dx \wedge dy}{xy^3}], [\frac{dx \wedge dy}{x^2y^2}],$$

$$[\frac{dx \wedge dy}{xy^4}], [\frac{dx \wedge dy}{x^2y^3}], [\frac{dx \wedge dy}{xy^5}], [\frac{dx \wedge dy}{x^2y^4}],$$

$$[\frac{dx \wedge dy}{xy^6}], [\frac{dx \wedge dy}{x^2y^5}],$$

and

$$([\frac{1}{x^2y^6}] - \frac{6b}{7}[\frac{1}{xy^7}] + \frac{2b^2}{7}[\frac{1}{x^3y}])dx \wedge dy.$$

One can easily see that $W_{Q_f} = \{0\}$. According to a result of K. Saito [29], we find that f is quasihomogeneous.

Remark Let $w = (7, 3)$ and let $\deg_w(\ )$ denote the weighted degree w.r.t. the weight vector w. Then $x^3 + y^7$ is a weighted homogeneous function of weighted degree 21. The weighted degree of the monomial xy^5 is 22 and that of the monomial xy^6 is 25. The monomials xy^5 and xy^6 are upper monomials. The monomial xy^5 satisfies the condition $\deg_w(xy^5) \leq 2 \times 21 - 2(7 + 3)$, whereas the monomial xy^6 does not satisfy the condition $\deg_w(xy^6) \leq 2 \times 21 - 2(7 + 3)$.

3. μ-constant deformations

Let $w = (w_1, w_2, ..., w_n) \in \mathbb{Z}^n$ be a weight vector and let f be a weighted homogeneous function (w.r.t. the weight vector w) with an isolated singularity at the origin $O \in X$. Let $\mu = \dim_{\mathbb{C}}(\mathcal{O}_{X,O}/J_f)$ denote the Milnor number of f.

Assume that a μ-constant deformation $f_t(x) = F(x,t), t \in T$ of f with $f_0(x) = f(x)$ is given, where $F(x,t)$ is a holomorphic function defined on $X \times T, X \subset \mathbb{C}^n$ and $T \subset \mathbb{C}^\ell$ are open neighbourhood of $O \in \mathbb{C}^n$ and of $O \in \mathbb{C}^\ell$.

Let $J_t = (\frac{\partial f_t}{\partial x_1}, \frac{\partial f_t}{\partial x_2}, \dots \frac{\partial f_t}{\partial x_n})$, the Jacobi ideal of f_t in the local ring $\mathcal{O}_{X,O}$. Then, the Milnor number $\mu_t = \dim_{\mathbb{C}}(\mathcal{O}_{X,O}/J_t)$ of f_t, which is independent of $t \in T$ from the assumption, is equal to μ.

We consider the following sets of parametric local cohomology classes defined by

$$W_{J_t} = \{\omega \in \mathcal{H}^n_{\{O\}}(\Omega^n_X) \mid \frac{\partial f_t}{\partial x_1}\omega = \frac{\partial f_t}{\partial x_2}\omega = \cdots = \frac{\partial f_t}{\partial x_n}\omega = 0\},$$

and

$$W_{T_t} = \{\omega \in \mathcal{H}^n_{\{O\}}(\Omega^n_X) \mid f_t\omega = \frac{\partial f_t}{\partial x_1}\omega = \frac{\partial f_t}{\partial x_2}\omega = \cdots = \frac{\partial f_t}{\partial x_n}\omega = 0\}.$$

Note that the dimension of the vector space W_{T_t}, which is equal to the Tjurina number $\tau_t = \dim_{\mathbb{C}}(\mathcal{O}_{X,O}/(f_t, J_t))$, depends on the parameter $t \in T$.

Let $\varphi_t : W_{J_t} \longrightarrow W_{J_t}$ be a map defined by $\varphi_t(\omega) = f_t\omega$. Set $W_{Q_t} = \mathrm{Im}\varphi_t$.

Lemma 3 *The sequence*

$$0 \longrightarrow W_{T_t} \longrightarrow W_{J_t} \longrightarrow W_{Q_t} \longrightarrow 0$$

of vector spaces defined by φ_t is exact.

Theorem 4

(i) $\dim W_{Q_t} = \mu - \tau_t$.

(ii) $Ann_{\mathcal{O}_{X,O}}(W_{Q_t}) = \{h \in \mathcal{O}_{X,O} \mid hf_t \in J_t\}$

Proof (ii) Since $W_{Q_t} = \mathrm{Im}\varphi_t$, we have

$$Ann_{\mathcal{O}_{X,O}}(W_{Q_t}) = \{h \in \mathcal{O}_{X,O} \mid h(f_t\omega) = 0, \; \forall \, \omega \in W_{J_t}\}$$
$$= \{h \in \mathcal{O}_{X,O} \mid (hf_t)\omega = 0, \; \forall \, \omega \in W_{J_t}\}.$$

Grothendieck local duality theorem implies

$$\mathcal{A}nn_{\mathcal{O}_{X,O}}(W_{J_t}) = J_t,$$

which also yields

$$\mathcal{A}nn_{\mathcal{O}_{X,O}}(W_{Q_t}) = \{h \in \mathcal{O}_{X,O} \mid hf_t \in J_t\}.$$

Example 3 (E_{18} singularity) Let $f(x,y) = x^3 + y^{10}$ and let

$$f_{(t_1,t_2)}(x,y) = F(x,y,t_1,t_2) = x^3 + y^{10} + t_1 xy^7 + t_2 xy^8.$$

The weight vector is $w = (10,3)$ and the Milnor number is equal to 18. The following 18 parametric local cohomology classes constitute a basis of the vector space W_{J_t}.

$$[\frac{dx \wedge dy}{xy}], [\frac{dx \wedge dy}{xy^2}], [\frac{dx \wedge dy}{xy^3}], [\frac{dx \wedge dy}{xy^4}], [\frac{dx \wedge dy}{xy^5}], [\frac{dx \wedge dy}{xy^6}], [\frac{dx \wedge dy}{xy^7}],$$

$$[\frac{dx \wedge dy}{x^2 y}], [\frac{dx \wedge dy}{x^2 y^2}], [\frac{dx \wedge dy}{x^2 y^3}], [\frac{dx \wedge dy}{x^2 y^4}], [\frac{dx \wedge dy}{x^2 y^5}], [\frac{dx \wedge dy}{x^2 y^6}],$$

$$([\frac{1}{xy^8}] - \frac{t_1}{3}[\frac{1}{x^3 y}])dx \wedge dy,$$

$$([\frac{1}{xy^9}] - \frac{t_1}{3}[\frac{1}{x^3 y^2}] - \frac{t_2}{3}[\frac{1}{x^3 y}])dx \wedge dy,$$

$$([\frac{1}{x^2 y^7}] - \frac{7t_1}{10}[\frac{1}{xy^{10}}] + \frac{7t_1^2}{30}[\frac{1}{x^3 y^3}] + \frac{7t_1 t_2}{30}[\frac{1}{x^3 y^2}])dx \wedge dy,$$

$$\omega_{(2,8)} = ([\frac{1}{x^2 y^8}] - \frac{t_1}{3}[\frac{1}{x^4 y}] - \frac{7t_1}{10}[\frac{1}{xy^{11}}] + \frac{7t_1^2}{30}[\frac{1}{x^3 y^4}] - \frac{4t_2}{5}[\frac{1}{xy^{10}}]$$

$$+ \frac{t_1 t_2}{2}[\frac{1}{x^3 y^3}] + \frac{4t_2^2}{15}[\frac{1}{x^3 y^2}])dx \wedge dy,$$

and

$$\omega_{(2,9)} = ([\frac{1}{x^2 y^9}] - \frac{t_1}{3}[\frac{1}{x^4 y^2}] - \frac{7t_1}{10}[\frac{1}{xy^{12}}] + \frac{7t_1^2}{30}[\frac{1}{x^3 y^5}] - \frac{t_2}{3}[\frac{1}{x^4 y}]$$

$$- \frac{4t_2}{5}[\frac{1}{xy^{11}}] + \frac{t_1 t_2}{2}[\frac{1}{x^3 y^4}] + \frac{4t_2^2}{15}[\frac{1}{x^3 y^3}])dx \wedge dy,$$

One can readily see that first 16 local cohomology classes in the list above belong to the space W_{T_t}. Therefore, from

$$\varphi_{(t_1,t_2)}(\omega_{(2,8)}) = -\frac{t_1}{30}[\frac{dx \wedge dy}{xy}],$$

$$\varphi_{(t_1,t_2)}(\omega_{(2,9)}) = (-\frac{t_1}{30}[\frac{1}{xy^2}] - \frac{2t_2}{15}[\frac{1}{xy}])dx \wedge dy,$$

we have

$$W_{Q_f} = \mathrm{Span}_{\mathbb{C}}\{t_1[\frac{1}{xy}]dx \wedge dy, (t_1[\frac{1}{xy^2}] + 4t_2[\frac{1}{xy}])dx \wedge dy\}.$$

4. An Example

We give an example to illustrate the proposed method to compute Tjurina stratification of the parameter space and the corresponding Tjurina numbers associated with a μ-constant deformation.

Let $f(x,y) = x^5 + xy^5$ and $w = (5,4)$. Set

$$f_{(t_1,t_2,t_3)}(x,y) = x^5 + xy^5 + t_1 x^3 y^3 + t_2 y^7 + t_3 y^8.$$

The Milnor number is equal to 21. We first compute, by using an algorithm described in [25], a basis local cohomology classes of the vector space W_{J_t}. We obtain the following 21 parametric local cohomology classes.

$$[\frac{dx \wedge dy}{xy}], [\frac{dx \wedge dy}{xy^2}], [\frac{dx \wedge dy}{xy^3}], [\frac{dx \wedge dy}{xy^4}], [\frac{dx \wedge dy}{xy^5}], [\frac{dx \wedge dy}{x^2 y}], [\frac{dx \wedge dy}{x^2 y^2}],$$

$$[\frac{dx \wedge dy}{x^2 y^3}], [\frac{dx \wedge dy}{x^2 y^4}], [\frac{dx \wedge dy}{x^3 y}], [\frac{dx \wedge dy}{x^3 y^2}], [\frac{dx \wedge dy}{x^3 y^3}], [\frac{dx \wedge dy}{x^4 y}], [\frac{dx \wedge dy}{x^4 y^2}],$$

$$([\frac{1}{x^5 y}] - 5[\frac{1}{xy^6}])dx \wedge dy, \quad ([\frac{1}{x^3 y^4}] - 3t_1[\frac{1}{xy^6}])dx \wedge dy,$$

$$([\frac{1}{x^4 y^3}] - \frac{3t_1}{5}[\frac{1}{x^2 y^5}])dx \wedge dy, \quad ([\frac{1}{x^5 y^2}] - 5[\frac{1}{xy^7}] + 7t_2[\frac{1}{x^2 y^5}])dx \wedge dy,$$

$$\omega_{(4,4)} = ([\frac{1}{x^4 y^4}] - \frac{12t_1}{25}[\frac{1}{x^6 y}] - \frac{3t_1}{5}[\frac{1}{x^2 y^6}])dx \wedge dy,$$

$$\omega_{(5,3)} = ([\frac{1}{x^5 y^3}] - 5[\frac{1}{xy^8}] - \frac{3t_1}{5}[\frac{1}{x^3 y^5}] + \frac{7t_2}{5}[\frac{1}{x^6 y}] - 7t_2[\frac{1}{x^2 y^6}]$$

$$+ (\frac{63}{25}t_1^2 t_2 + 8t_3)[\frac{1}{x^2 y^6}])dx \wedge dy,$$

and

$$\omega_{(5,4)} = \left(\left[\frac{1}{x^5 y^4}\right] - 5\left[\frac{1}{xy^9}\right] - \frac{12t_1}{25}\left[\frac{1}{x^7 y}\right] - \frac{3t_1}{5}\left[\frac{1}{x^3 y^6}\right] - \frac{7t_2}{5}\left[\frac{1}{x^6 y^2}\right] - 7t_2\left[\frac{1}{x^2 y^7}\right]\right.$$

$$+ \frac{9t_1^2}{5}\left[\frac{1}{xy^8}\right] - \frac{49t_2^2}{5}\left[\frac{1}{x^3 y^5}\right] + \left(\frac{63}{125}t_1^2 t_2 - \frac{8}{5}t_3\right)\left[\frac{1}{x^6 y}\right]$$

$$+ \left(\frac{63}{25}t_1^2 t_2 + 8t_3\right)\left[\frac{1}{x^2 y^6}\right] - \frac{147}{5}t_1 t_2^2\left[\frac{1}{xy^7}\right]$$

$$\left. + \left(-\frac{72}{25}t_1^2 t_2 - \frac{1029}{25}t_1 t_2^3\right)\left[\frac{1}{x^2 y^4}\right]\right)dx \wedge dy.$$

It is easy to see that the first 18 cohomology classes in the list above belong to the space W_{T_t}. In order to obtain a set of generators of the vector space W_{Q_t}, we compute three local cohomology classes $\varphi_t((\omega_{(4,4)})$, $\varphi_t(\omega_{(5,3)})$ and $\varphi_t(\omega_{(5,4)})$. From

$$\varphi_t(\omega_{(4,4)}) = \left(-\frac{2}{25}t_1\left[\frac{1}{xy}\right]\right)dx \wedge dy, \quad \varphi_t(\omega_{(5,3)}) = \left(\frac{3}{5}t_2\left[\frac{1}{xy}\right]\right)dx \wedge dy,$$

and

$$\varphi_t(\omega_{(5,4)}) = \left(-\frac{2}{25}t_1\left[\frac{1}{x^2 y}\right] + \frac{3}{5}t_2\left[\frac{1}{xy^2}\right] + \left(-\frac{27}{125}t_1^2 t_2 + \frac{7}{5}t_3\right)\left[\frac{1}{xy}\right]\right)dx \wedge dy,$$

we find that the space W_{Q_t} is spanned by

$$t_1\left[\frac{1}{xy}\right]dx \wedge dy, \ t_2\left[\frac{1}{xy}\right]dx \wedge dy, \ \left(-2t_1\left[\frac{1}{x^2 y}\right] + 15t_2\left[\frac{1}{xy^2}\right] + 35t_3\left[\frac{1}{xy}\right]\right)dx \wedge dy.$$

By using a parametric version of the standard basis computation presented in [35], we obtain the following stratification.

(0) if $t_1 = t_2 = t_3 = 0$, then $\mu - \tau = 0$,

$$W_{Q_0} = \{0\}, \ f \in J.$$

(1) if $t_1 = t_2 = 0, t_3 \neq 0$, then $\mu - \tau_t = 1$,

$$W_{Q_t} = \text{Span}\left\{\left[\frac{dx \wedge dy}{xy}\right]\right\}, \ J_t : f_t = (x, y).$$

(2) if $t_1 \neq 0$, or $t_2 \neq 0$, then $\mu - \tau_t = 2$.

(i) if $t_1 = 0, t_2 \neq 0$, then

$$W_{Q_t} = \text{Span}\left\{\left[\frac{dx \wedge dy}{xy}\right], \left[\frac{dx \wedge dy}{xy^2}\right]\right\}, \ J_t : f_t = (x, y^2).$$

(ii) if $t_1 \neq 0, t_2 = 0$, then

$$W_{Q_t} = \text{Span}\{[\frac{dx \wedge dy}{xy}], [\frac{dx \wedge dy}{x^2 y}]\}, \quad J_t : f_t = (x^2, y).$$

(iii) if $t_1 \neq 0, t_2 \neq 0$, then

$$W_{Q_t} = \text{Span}\{[\frac{dx \wedge dy}{xy}], (-2t_1[\frac{1}{x^2 y}] + 15t_2[\frac{1}{xy^2}])dx \wedge dy\},$$

and

$$J_t : f_t = (2t_1 y + 15t_2 x, xy, y^2).$$

Some remarks are in order.

Remarks

(1) The existing algorithm due to B. Martin and G. Pfister [22] is based on the deformation theory and utilize the Kodaira-Spencer maps. The proposed method is free from the deformation theory.

(2) The proposed method can be extendable to handle the case where even the quasi-homogeneous part contain parameters.

(3) The method is also applicable to compute the Hilbert function of the Tjurina algebra $\mathcal{O}_{X,O}/(f_t, J_t)$ (cf. [7]).

References

1. V. I. Arnold, S. M. Gusein-Zade and A. N. Varchenko, *Singularities of Differentiabla Maps*, I, Birkhauser 1985.
2. C. Bănică et O. Stănăşilă, *Méthodes Algebriques dans la Théorie Globale des Espaces Complexes*, Gauthier-Villars, 1974.
3. J. Briançon, M. Granger and Ph. Maisonobe, Le nombre de modules du germe de courbe plane $x^a + y^b = 0$, *Math. Ann.* **279** (1988), 535–551.
4. P. Cassou-Noguès, Etude du comportement du polynôme de Bernstein lors d'une deformation à μ-constant de $x^a + y^b$ avec $(a, b) = 1$, *Ann. Inst. Fourier, Grenoble* **36** (1986), 1–30.
5. A. M. Dickenstein and C. Sessa, Duality methods for the membership problem, *Progress in Math.* **94** (1991), Effective Methods in Algebraic Geometry, 89–103, Birkhäuser.
6. G. -M. Greuel and G. Pfister, On moduli space of semiquasi-homogeneous singularities, *Progress in Math.* **134** (1996), 171–185, Birkhäuser.
7. G. -M. Greuel and C. Hertling, Moduli spaces of semiquasihomogeneous singularities with fixed principal part, *J. Algebraic Geom.* **6** (1997), 169–199.
8. P. Griffiths and J. Harris, *Principles of Algebraic Geometry*, Wiley-Interscience Pub. 1976.

9. A. Grothendieck, *Théorèmes de dualité pour les faisceaux algébriques cohérents*, Séminaire Bourbaki **149** (1957).

10. A. Grothendieck, *Local Cohomology*, Lecture Notes in Math. **41** (1967), Springer.

11. R. Hartshorne, *Residues and Duality*, Lecture Notes in Math. **20** (1966), Springer.

12. C. Hertling and C. Stahlke, Bernstein polynomial and Tjurina number, *Geometriae Dedicata* **75** (1999), 137–176.

13. M. Kashiwara, On the maximally overdetermined system of linear differential equations, I, *Publ. Res. Inst. Math. Sci.* **10** (1975), 563–579.

14. H. Komatsu, Projective and injective limits of weakly compact sequences of locally convex spaces, *J. Math. Soc. Japan* **19** (1967), 366–383.

15. H. Komatsu, Relative cohomology of sheaves of solutions of differential equations, *Lecture Notes in Math.* **287** (1973), 192–261.

16. E. Kunz, *Residues and Duality for Projective Algebraic Varieties*, AMS. 2008.

17. O. A. Laudal and G. Pfister, *Local Moduli and Singularities*, Lecture Notes in Math. **1310**, 1988.

18. O. A. Laudal, B. Martin and G. Pfister, Moduli of plane curve singularities with C^*-action, *Banach Center Pub.* **20** (1988), 255–278.

19. J. Lipman, *Dualizing sheaves, Differentials and Residues on Algebraic Varieties*, Astérisque **117** (1984)

20. Lê Dung Trang and C. P. Ramanujan, The invariance of Milnor number implies the invariance of the topological type, *Amer. J. Math.* **98** (1973), 67–78.

21. M. G. Marinari, H. M. Möller and T. Mora, Gröbner bases of ideals given by dual bases, *ISSAC* 1991, ACM (1991), 55–63.

22. B. Martin and G. Pfister, The kernel of the Kodaira-Spencer map of the versal μ-constant deformation of an irreducible plane curve with C^*-action, *J. Symbolic Comp.* **7** (1989), 527–531.

23. M. Morimoto, *An Introduction to Sato's Hyperfunctions*, Translations of Mathematical Monographs, **129** (1993), AMS.

24. B. Mourrain, Isolated points, duality and residues, *J. Pure and Appl. Alg.* **117 & 118** (1997), 469–493.

25. K. Nabeshima and S. Tajima, On the computation of algebraic local cohomology classes associated with semi-quasihomogeneous singularities, to appear in *Adv. Studies in Pure Math.*

26. Y. Nakamura and S. Tajima, On weighted-degrees for algebraic local cohomologies associated with semiquasihomogeneous singularities, *Adv. Studies in Pure Math.*, **46** (2007), 105–117.

27. M. Noro and T. Takeshima, Risa/Asir - a computer algebra system, *ISSAC* 1992 (ed. P. S. Wang), ACM (1992), 387–396.

28. R. Peraire, Tjurina number of a generic irreducible curve singularity, *J. Algebra* **196** (1997), 114–157.

29. K. Saito, Quasihomogene isolierte Singularitaten von Hyperflächen, *Invent. Math.* **14** (1971), 123–142.

30. M. Sato, Theory of hypersunctions, II, *J. Fac. Sci. Univ. Tokyo* Sect. I **8** (1960), 387–437.

31. G. Sorani, Sulle rapprezentazione delle funzioni olomorfe, *Atti. Accad. Naz. Lincei Rend. Cl. Sci. Fiz. Mat. Natur.* **39** (1965), 161–166.

32. S. Tajima, Methods for computing zero-dimensional algebraic local cohomology and standard bases computation (in Japanese) *RIMS Kôkyûroku* **1514** (2005), Kyoto, Computer Algebra–Design of Algorithms, Implementations and Applications, 126–132.

33. S. Tajima and Y. Nakamura, Algebraic local cohomology classes attached to quasi-homogeneous isolated hypersurface singularities, *Publ. Res. Inst. Math. Sci.* **41** (2005), 1–10.

34. S. Tajima and Y. Nakamura, Annihilating ideals for an algebraic local cohomology class, *J. Symbolic. Comp.* **44** (2009), 435–448.

35. S. Tajima, Y. Nakamura and K. Nabeshima, Standard bases and algebraic local cohomology for zero dimensional ideals, *Adv. Studies in Pure Math.*, **56** (2009), 341–361.

36. S. Tajima and Y. Nakamura : Algebraic local cohomology classes attached to unimodal singularities, *Publ. Res. Inst. Math. Sci.*, **48** (2012), 21-43.

37. S. Tajima, On polar varieties, logarithmic vector fields and holonomic D-modules, to appear in *RIMS Kôkyûroku Bessatsu*.

38. F. Treves, *Topological Vector Spaces, Distributions and Kernels*, Academic Press, 1967.

39. A. N. Varchenko, A lower bound for the codimension of the μ=constant in terms of the mixed Hodge structure, *Moscow Univ. Math. Bull.* **37** (1982), 30–33.

Author Index